PROSPERITY, FREEDOM, AND ORDER

A DURABLE POLITICAL ECONOMY AND A FREE ORDERLY SOCIETY

BY

THE INSTITUTE OF PERCEPTIONISM® INC.

Prosperity, Freedom, and Order: A Durable Political Economy and a Free-Orderly Society

For more information, visit: https://theinstituteofperceptionism.com/

First Edition: Published in 2014 by The Institute of Perceptionism, Inc.

Second Edition: Published in 2025 by Richard Allen Beach

Library of Congress Control Number: 2014937395

ISBN Paperback: 978-1-969570-09-4

ISBN Hardcover: 978-1-969570-10-0

Other Writings of the Institute of Perceptionism®

Other writings offered by the Institute of Perceptionism, Inc. include:

- Proposed Amendments
- Healthcare
- The Four Pillars of Civilization
- Sport Team Ownership

2025
Second Edition

TABLE OF CONTENTS

PREFACE

This book provides the reader with what, to this point, is a unique perspective of economics. It offers several new economic institutions that will create a stable, growing economy, avoiding booms and busts; make the whole of society ever more prosperous and effective; lessen the great disparity of wealth and income; and aid economic recovery and transition. It offers the reader a clear, irrefutable understanding of individual freedom and an undeniable code of behavior in a free, orderly society. For the reader having an open mind, a view of a wonderful world awaits you. Here we try to describe how this point of view of our world impacts the four pillars of a civilization—the nuclear family, religion, government, and economic system—as well as recognize an expansion of the categories of basic human need from just food, clothing, and shelter to include healthcare, education, transportation, sanitation, and communication.

Prosperity, Freedom, and Order offers an economic, social, and moral structure that provides prosperity and individual freedom and order that is compatible with most traditional religions and beliefs, so they can be practiced in peace and harmony.

PART ONE
A FREE AND ORDERLY SOCIETY

CHAPTER 1

THE NATURE AND VALUE OF FREEDOM

The Equilibrium Economy presented above is essential to a free and orderly society. Here, we establish the basics for the reader, avoiding intricacies and nuances. It is upon this basic framework that one can build an understanding of the nature of freedom. Subtleties will be developed further on.

"Natural Freedom" or "Generic Freedom"

A free society is one having an ***absence of coercion***. The antithesis of freedom is a state of complete coercion—a state of slavery or absolute tyranny.

Freedom, as used here, is not a "granted" liberty. It is not a license. It is not a privilege granted by a higher authority. *Freedom is a state of being where there exists an absence of coercion by one person over another or by government over the citizens.*

To understand natural freedom or generic freedom, one must be able to distinguish between societal restriction and coercion, and personal restriction and coercion. Too frequently, these are confused, and attempts are made to have society responsible for freeing individuals from their personal restrictions and personal coercions.

Individuals are restricted by their natural limitations. Their sight is poorer than that of the birds, and some individuals' sight is poorer than others. While they accept the fact that their sight is poorer than the birds', they sometimes feel that it is a societal responsibility to assist them with their disability as they relate to other individuals. The same can be said about their hearing as compared to other species or other individuals. Or, it can be said of their mobility as compared to other species or individuals. All individuals have personal restrictions. These are not "societal restrictions" and cannot be ameliorated by government in a free society. To do so is to arbitrarily use the power of government to coerce other citizens to participate in the amelioration. This is, or would be by any definition, a denial of their individual freedom. This enslaves, if only in part, one portion of the population into the service of another.

Similarly, individuals are coerced by their very nature, by their natural environment, and by their personal Needs. The word "Needs," as used in this writing, refers to those human motivators such as hunger, sex, ego, and the like that are responsible for all human activity and that constantly coerce each individual. But these are not societal coercions, nor can they be made so in a free society. Individuals who have a desire or need to help others meet such Needs can do so, but the government cannot shift responsibility for meeting such Needs to other citizens of a free society. That could only be accomplished by using the coercive power of government. Also, individuals are coerced by nature—by floods, earthquakes, snow, and rainstorms. Here again, in a free society, government cannot use its coercive power to provide relief from these natural coercions. In a free society, insurance and voluntary private organizations must be relied on to provide relief, for any other method must employ the coercive power of government. Such is incompatible with individual freedom. This is not to suggest that a government committed to culturing, producing, and preserving a free, orderly society cannot

create or charter private not-for-profit institutions that would provide relief from natural disasters. As will be advanced further on, natural disasters that threaten the survival of humanity or the free-orderly society fall within the purview of defense and preservation of the society. Such calamities should be addressed by the government.

For a population that has looked to government for solutions, relief, care, and compassion, voluntary private problem-solving is a difficult concept to comprehend. In the United States, there was a movement from self-reliance that prevailed at the end of the 19th century to an attitude, a century later, of shifting responsibilities from the individual, the family, and the church to the government and thereby to other citizens. To be a free society, individuals must become competent and self-reliant. They will seek to provide for their own risks through avoidance of exposure to the risk, insurance, and joining organizations dedicated to providing help in such events. Examples are volunteer fire departments and community relief groups. Those who are seriously dedicated to individual freedom will look to and contribute to private agencies that evaluate and ameliorate risk in all conceivable areas. They will look to and participate in all manner of organizations that provide care and compassion. They will seek educational institutions that will teach their children self-reliance and the acceptance of responsibility for one's, one's family, particularly one's minor children, and one's property. Failure to proceed in such a direction is to abandon any serious commitment to individual freedom. This will become very clear as you proceed with this writing.

Societal Freedom or the Sovereignty of the Individual versus Liberty

We tend to equate "liberty" and "freedom." However, they are quite different. Our founders wanted liberation from the sovereignty of the King and his government. Once liberated, sovereignty rested with the individual—now that each individual was equally sovereign to a king and thereby implicitly equally sovereign to each other, "all men are created equal."

Freedom of the individual can be defined in a social sense as the sovereignty of the individual. It is widely understood that a totally sovereign king or dictator has life and death authority over his subjects. It is not so widely understood that the sovereign individual has total authority over his/her person, property, and minor children. But it must not be overlooked that the sovereign individual has total responsibility for his/her self, property, and minor children. *The sovereign individual is a free individual, and it is here where we are all equal.* To whatever degree the individual relinquishes his/her sovereignty, he/she loses his/her freedom. Sovereign individuals can be expected to relinquish their sovereignty to government, commonly spoken of as "the rules of the road"; however, they must be determined to be *essential to preserving the remainder of their sovereignty and an orderly interfacing of individuals.*

Many individuals have a sense that they are never sovereign but rather a subject of a higher being—God. These persons must recognize that other individuals may feel subject to a different higher being, while still others feel subject to no higher being. It is for this reason that no one's religious moral code can be forcibly imposed upon another in a free-orderly society. The sovereign individual is free to subjugate him/herself to whatever or whomever he/she chooses. But in no way can one expect to subjugate other sovereign individuals to his/her religious code or value structure in a free-orderly society.

Authority and Responsibility

Where authority over another exists, the freedom of the subject is diminished. This concept addresses the very essence of freedom. *Freedom is a state where there exists an absence of coercion of one*

person by another and of citizens by their government. Where an individual has authority over another, this authority only exists if it can be enforced. If one individual asserts authority over an unwilling individual, you have the coercion of slavery. When the government exerts authority and asserts that authority over unwilling citizens, you have tyranny.

The government cannot accept responsibility; it can only shift responsibility. It does so, generally, by exercising its coercive taxing powers. But it can also exercise authority directly through requirements and prohibitions punishable by law. Explicitly or implicitly, it exercises such authority because it has assumed—shifted—consequential responsibility.

For example, if society were to be responsible for the head injuries of motorcycle riders, it assumes the authority to require them to wear helmets. If society assumes—shifts—responsibility for the health of drug addicts, it assumes authority to prohibit drug use. If society assumes—shifts—responsibility for the diseases resulting from obesity, it is implicitly granted authority over what and how much people can eat. It does so only at the expense of the freedom of its citizenry.

It is axiomatic, therefore, that to the extent that government uses its coercive powers to shift responsibility to some of its citizens, it is partially enslaving those citizens to the service of others.

Wherever and whenever the government exercises authority over its citizens, it diminishes freedom.

Within the framework of behaving in a manner that does no harm to the person, property, and minor children of all others, if one were to be totally free, one would exercise total authority over his/her person, property, and minor children. In a free-orderly society, individuals must be totally responsible for their own actions and those of their minor children and their animals (property). Sovereign individuals implicitly have authority over their activities, those of their minor children, and control of their property. But they must be held responsible for all of the consequences of all of their actions and those of their minor children and their animals, and other property as they relate to others. They have, by their actions, exercised their authority and, commensurately, must be held responsible. There is no mechanism in a free-orderly society for shifting the responsibility for one's actions to another who does not willingly accept them. Therefore, in a free-orderly society, individuals cannot act in an irresponsible manner and shift responsibility to others who would otherwise be unwilling to accept the responsibility.

Some would equate freedom with freedom from accepting responsibility. That could only be in a free-orderly society if someone else would willingly accept the responsibilities of another, such as parents accepting responsibility for an adult child. One can easily observe that prisoners have virtually no responsibilities, for almost all of their responsibilities have been shifted to the taxpayers. They are free of responsibilities, but they are anything but free.

If one has a problem that is not a result of someone else's direct action, it is that person's problem and no one else's if there is no one else willing to accept that problem voluntarily. Any physical or mental inadequacy or infirmity, if a problem, remains a problem of that individual. Responsibility for such personal problems cannot be shifted to others unwilling to accept them in a free-orderly society. This is not to suggest that there will not be others willing to accept responsibility and voluntarily assist those in need. Responsibility for financial obligations resulting from all manner of circumstances may be shifted to insurance companies that willingly accept the financial responsibility through a contract.

An understanding of these concepts is essential to a comprehension of freedom. In the concepts that follow, dealing with government, family, and religion, "authority, responsibility, and freedom" will be developed in each context.

By the very nature of freedom, authority, and responsibility, one's authority is limited to one's self, one's property, and one's minor children, and any other persons who willingly grant their authority to him/her. Equally, in a free-orderly society, one's responsibility is no greater or lesser than that over which he/she has willingly accepted total authority. Where one grants authority over his/her self to another, such as an employer, commensurate responsibility is conferred with that authority. For example, when a roofing company employs a roofer, the roofer has granted authority over him/herself to the company. The company, now having the authority to instruct the person to climb on a roof, accepts the responsibility for any injuries that the person may incur while following the employer's instructions. Where the military has authority and command over the military personnel, the government has the responsibility for the care of the injured and their dependents. A person serving in the military under a contract cannot be held responsible for any acts committed during such employment beyond those expressly required or prohibited in the enlistment contract. A conscripted military person cannot be held accountable for anything, for he/she serves without consent. Having not consented to accepting responsibility, all acts committed during such service become the responsibility of the employer, the military. Responsibilities resulting from accepting rank must be expressly acknowledged.

In the case of a police force, individuals are employed by governments to maintain a free-orderly society. These individuals are hired to act as agents of their employer. As agents, acting under the terms of their employment contract, the employer is responsible for any and all acts committed in their efforts to fulfill the terms of their employment in good faith. Any failure or mistake committed while on duty remains the responsibility of the employer—government entity. Therefore, all police officers must have total indemnification while on duty. Failure to be provided with such is a gross misjudgment in the selection of an occupation.

Assigning responsibility to an individual without authorizing commensurate authority to execute that responsibility is, on its face, untenable. For example, assigning the responsibility to turn off the lights in a building without giving the person the keys to enter the building and gain access to the switches is untenable.

These concepts underlie the very foundation of a free-orderly society. To ignore or to fail to comprehend it is to fail to comprehend the fundamental nature of individual freedom. To deny the preceding concepts is to deny an acceptance of free individuals living in a free-orderly society.

Freedom, therefore, can be no less than the sovereignty of the individual over his/her person, property, and minor children. This is in regard to all adult individuals regardless of gender, race, or any other consideration. It is here, and only here, that we are all equal. And this is the lowest and most basic common denominator of social organization.

Relative Value of Human Life

Whenever and wherever human life was more precious than freedom, tyrannous repression, enslavement, and serfdom prevailed. Whatever freedom exists, it only exists because some people were willing to place their lives at risk to achieve it, preserve it, and defend it. The citizen army that placed the survival of Athens above their own lives earned the freedom experience for the Athenians. This has been repeated many times since the Battle of Marathon in 490 BC.

Rome's citizen armies won the freedom that was the product of the Roman Republic. The Minute Men of the American Revolution are most certainly in that classification. During the twentieth century, hundreds of thousands died in the wars that thwarted the tyrannous ambitions of Nazi Germany, Imperial Japan, and the wars that contained tyrannous Marxism until its collapse. If one were to conclude that life is more precious than freedom, one would have to conclude that those who lost their lives fighting the expansion of the tyrannies of the Nazis, the imperialists, and the Communists were stupid and misguided. Freedom is not naturally occurring in human social organization.

The consequences of subordinating freedom to human life have produced enslavement in many forms. Most notably, the "democratic socialist" governments have evolved a creeping tyranny where freedom of the individual becomes endlessly diminished. As individual freedom never becomes of greater importance than the preservation and protection of each human life, governments have required greater and greater amounts of resources to be allocated to sustaining and protecting the lives of their citizens. No amount of resources allocated will be enough. All will die regardless of the resources spent. The ever-increasing allocation of resources to sustaining lives means continuous diminution of individual freedom. In such a society, resources allocated to raising the standard of living through the accumulation of capital assets, to scientific research, and exploration have become ever diminishing, as they, along with individual freedom, are of lower priority than the preservation of each human life.

Freedom has been virtually unknown to most societies through nearly all of recorded history. With only the briefest exceptions affecting only the smallest portion of the population, the whole human experience having any semblance of freedom has occurred, and most of this has been in the last half of the twentieth century. Fifty years, compared to the thousands that preceded it, should be sufficient evidence to suggest that freedom, as a social order for human life, is not its ordinary form. One should conclude that freedom is a rare social order, and is fragile and easily compromised. Freedom is so conducive to high productivity and rapid progress that it should be treated as humanity's most precious possession.

Human life and freedom cannot be of equal importance. *If individual freedom is not more important than life, there will be no individual freedom.* As has been pointed out, freedom exists only because millions were and are willing to fight and die for it. Where life is deemed more important than freedom, freedom is ever diminished and vanishing. In the United States, there is a pervasive problem of understanding the conflict between freedom and human life. Virtually all schoolchildren know that the Declaration of Independence endows each with unalienable rights, including life, liberty, and the pursuit of happiness. Only months before that declaration, Patrick Henry, in his declaration "Give me liberty or give me death," stated the potential conflict between liberty and life. (At that time, individual freedom was a grant from an authoritative government that properly described it as a liberty.) Very frequently, there is a distinct conflict between life and freedom. When there is, each cannot be of equal weight. One must yield to the other.

Populations generally recognize that when freedom is threatened from without, freedom is more precious than life. But they fail to recognize that priority when institutions of special interests threaten freedom from within. The word "life," as used here, includes by its very nature all manner of care and compassion without which life is shortened. In Western societies, freedom has yielded to the care, compassion, and preservation of life.

Government cannot confiscate the production and assets of one citizen in order to preserve the life of a murderer; to extend the life of the aged; to save the life of an infant or newborn, and not honor the fact that it is enslavement. Regardless of who or how the slave's production is used, it is still slavery. All such acts grant life a higher rank than freedom.

Yielding to the subordination of freedom to life has resulted in a continuous erosion of freedom, with there being no discernible limit until personal freedom has been extinguished. To reiterate this crucial point, there is no limit to the amount of resources that can be allocated to the preservation of life, and as a result, there is no limit to the extent that human resources can be conscripted to that end. Society must be concerned with the survival of society, which is different from concern for the survival of any individual.

One who bothers to look at advanced twenty-first-century societies will see the dominance of life over freedom in every corner of government involvement in human activities. This is the tyranny of government, regardless of whether it is a democracy or a dictatorship. One cannot seriously consider freedom and at the same time dismiss the constant conflict between freedom and life. This will be more apparent as one proceeds through this writing.

An orderly society is essential to individual freedom.

Chaos breeds tyranny. *History teaches that societies confronted with the choice of chaos or tyranny will choose tyranny.* An economy of specialization and the division of labor requires an orderly society. It requires respected property rights and a reliable currency, one accepted in free and orderly economic intercourse. For without this specialization and division of labor, the productivity of individuals and productivity in the aggregate can decline to below subsistence. *A society cannot be free without being orderly. Therefore, any human life is not more precious than a free-orderly society.*

A society that mandates order must assign order a higher priority than life. It follows, therefore, that *society must restrict freedom only to the extent necessary to require that all who live in a free society behave in a manner that does no harm to the person, property, or minor children of others.* Such behavior will produce an orderly society.

Life is not so precious as to preserve the life of a murderer at the expense of the freedom of law-abiding citizens. A free-orderly society will not confiscate the assets and income of its law-abiding citizens to maintain a murderer in prison. This example, along with others, will be developed further.

To further the understanding that "while life is precious, it is not more precious than everything else," we need only to view the conduct and judgment of individuals as they conduct their lives. Life is very precious, but not so precious that individuals do not risk their lives daily to travel the necessary distance to and from their place of employment in order to achieve that which, to them, is a more desirable place to reside. Nor is life so precious as to prevent persons from engaging in occupations that, by their nature, pose a significant risk to life. Life is very precious, but not so precious that individuals do not risk their lives to travel vast distances in and on all kinds of conveyances, pursuing gratification of their curiosity or other motivations instead of remaining in the greater safety of their homes. Life is precious, but not so precious that individuals do not risk their lives in all manner of travel, thrill-seeking, and other activities. So, one can conclude that travel in the aggregate is more valuable than life. Life is indeed precious, but it is not more precious than everything else. It is not more precious to individuals than the right to freely pursue a full range of life's activities in a free-orderly society. No aspect of a free-orderly society can be compromised for the preservation of life without the compromises becoming endless and the free-orderly society being destroyed.

Viewing recorded history, one might reasonably conclude that a free-orderly society would be the highest social achievement of mankind. It is a fact demonstrated clearly time and again throughout history to anyone willing to see: totalitarian societies restrict creativity, they destroy the search for truth, and they fall prey to the rankest corruption and end up making the human condition deplorable. Democracy provides little or no protection against becoming a tyrannous bureaucracy. Some have equated democracy with free-orderly societies. Such thinking is erroneous. Democratic government must not be confused with a free-orderly society. *Democratic governments have political parties that are coalitions of conspirators that use the coercive power of government to impose their will on others, and they thereby fail the test of free-orderly societies.*

Value of Freedom

Freedom makes life more pleasant for most who live in it. But its value is to be found in what it can produce beyond the hedonistic. In part, freedom permits human beings to pursue their religious and spiritual beliefs and thereby find fulfillment and happiness that might otherwise be precluded in tyrannized societies. The experience of those who were subjugated to socialism in Eastern Europe in the twentieth century gives stark proof of this fact. It is the individuals' pursuit of human Needs and their fulfillment that makes them productive. It is this productivity that enables humanity to produce its great achievements.

Societies that have denied freedom of thought have stagnated for hundreds of years. Societies that have nurtured freedom of thought have experienced the most rapid advancement and the greatest achievements. Billions of brains, having billions of combinations of thought processes, if free to think and act, will maximize the creative potential of humanity. *Freedom maximizes the achievements of humankind.* Although most of this free thinking will be creatively unproductive, the greatest possible total of productive creative thinking will develop in a free society. No one person, or group of humans, can judge which creative thoughts will be productive. By its very nature, there is no scale or precedent upon which to evaluate innovations, creative thoughts, or concepts. Only the test of application and time will determine the productive merit of any particular creative thought. Human thought that was dominated by a set of precepts retarded productive Western thinking for a thousand years, essentially from the fifth to the fifteenth century in Europe. The very nature of the process of scientific exploration is, itself, creative thinking. Here again, the words of Arthur Schopenhauer are helpful: "Talent can hit a target that no one else can hit; genius can hit a target that no one else can see."

A free society is the most productive in terms of serving human Needs and wants. Its by-product is the strongest possible economic system. The developed countries having the greatest total wealth in the twentieth and twenty-first centuries have allocated large sums of money to scientific research and exploration. Those sums are but a very small part of the total production of these countries. The great space explorations are among the most visible achievements of humanity, but vast sums are allocated to medical research and other less visible fields of research. Poverty-stricken societies produce extremely little in these fields.

There is no known maximum speed or rate at which humanity can expand its knowledge. The rate of acquisition of knowledge has varied widely from society to society and from time to time. Experience shows us that the acquisition of knowledge can only be maximized in a social order of freedom of individual action and thought. Purpose is also to be found in these pursuits.

CHAPTER 2

THE MORALITY OF A FREE-ORDERLY SOCIETY

A free-orderly society has an overriding moral code by which it can judge all human behavior. It is simple, easily understood, unambiguous, and a universal standard of conduct that can be practiced by all. The experience of many societies over millennia has demonstrated that complex moral codes, filled with implicit contradictions, are confusing and, as a result, ineffectual. Commonly, those who govern by such codes exempt themselves from an obligation to respect the codes—if not totally, at least to some degree. Having an ineffective moral code is like having no moral code.

Inasmuch as moral codes have had religious beliefs as their foundation, and a free society by its very nature accepts the practice of many religions, then implicit in such a structure, there exist many conflicting moral views about many different issues. In a society that has any degree of democratic government or representative government, there results an imposition of moral values acceptable to one group that is unacceptable or even intolerable to another group. Such a hodgepodge leaves such a society governed by a moral muddle.

A true moral code, by definition, could and would be acceptable to all. Therefore, all persons in such a society could practice their own moral code, pursuing their own moral values and virtues, without interference from others. And so, it undeniably follows that each person must, when practicing his/her own moral code, do so without interfering with any other person. This true moral code enables each person to engage in infinite personal pursuits, following an infinite number of value structures, accommodating an infinite number of personal situations and Needs.

If the principal objective of a society is to be a free-orderly society, a moral code can be easily constructed by heeding the natural principles of freedom of the individual and the essential elements of his/her relationship to all others with whom contact is made.

Respect for the physical and mental well-being of others is fundamental to a free-orderly society. Society, therefore, cannot compromise the requirement that all of its participants conduct themselves in a manner that does not harm other persons in it. That statement is simple and easily understood. To say it another way would be that the participants must do no harm to any other person. This does not impose any obligation on one person for the well-being of another. Such obligations may only be personally accepted by an individual or imposed on an individual by their own religion, in a free-orderly society.

To be a free individual requires that the individual must be sovereign (have total authority and inseparably total responsibility) over his/her person, property, and minor children. It is not possible for an individual to be free without being sovereign over his/her person. This will be developed in more depth further on.

Respect for the property of others and their sovereignty over it is essential to a free-orderly society. Society, therefore, cannot compromise the requirement that all of its participants conduct themselves in a manner that respects an individual's sovereignty over their property by all others and does no harm to the property of others. Here too, this statement is simple and easily understood. It can be universally applied and is, therefore, by definition, a true morality.

Private ownership of property or of the product of the individual is an essential element of a free society. This has been explained by scholars over the centuries and thoroughly established by human experience. It will be further developed below.

Minor children are such because the human child is incapable of caring for itself. This commences at birth with near total incompetence to some arbitrary age that society grants competence adulthood a point in life when an individual can fend for himself/herself. Understanding the inseparability of responsibility and authority in a free-orderly society, the parents must have both total responsibility and total authority until both are granted to and vested in the minor, making him/her an adult. To not harm the minor children of others and respect the sovereignty of the parent over them is essential to a free-orderly society. This is a difficult concept for many to grasp, but the understanding of it is paramount to a free-orderly society.

To help understanding of this, it can be asked, "If one lived in South Carolina in 1800 and were black, would she/he be sovereign over his/her person, his/her property—production—or minor children?" The answer would be, "No matter how one evaluates it, sovereignty over any of these three elements did not exist for the black individual!" It was denied by the coercive power of the government.

The very existence of a free-orderly society requires that all of its participants behave in a manner that does no harm to the person, property, and minor children of all others and honors their sovereignty over those. There is, therefore, no choice but to require that all of its citizens and those in its jurisdiction behave in such a manner. For those persons who choose not to behave in such a manner, or are not capable of such behavior, a free-orderly society requires their expulsion or banishment. Like it or not, there is no alternative.

The success of any person requires that they achieve these abilities and inhibitions. From a child's earliest experiences, they must learn not to injure or do harm to others. This, of course, will include not hitting or telling lies that would harm others. They must learn not to harm, take, or use the property of others without permission.

Any person who demonstrates an inability (after having had the opportunity to modify his/her behavior patterns) to not harm the person, property, or minor children of others must be removed, banished, or otherwise expelled from the free-orderly society. To compromise this principle is to deny it and to assure the failure of a free-orderly society. Any other solution would require the government to impinge upon the property of the citizenry through taxation in order to imprison such people or to tolerate their harming the person, property, or minor children of other citizens.

Sigmund Freud (1856–1939) expressed concern that innate aggression in the human psyche will always become manifest. In a free-orderly society, these, if insurmountable, must be channeled in ways that do no harm to the person, property, or minor children of others, or to oneself, one's property, or one's minor children. Perhaps the release of these aggressions is successfully achieved in socially acceptable ways through participation in or viewing competitive games and sports. Not having the ability or willingness to control such aggressions in a manner that does not harm the person, property, and minor children of others will cause the person to be expelled, by whatever method, from a free-orderly society.

A society having an uncompromising respect for private property cannot sanction a government that does not. Government cannot morally confiscate the production or property of one individual, using the coercive power of its armed agents, and then give it to or use it for the benefit of another. That has no more moral foundation than for the recipient to have confiscated the same production or assets at gunpoint for his own benefit. Both fail any test of respecting property rights.

The use of force to confiscate the property that serves to benefit one person, one family, one region, or one nation to be used for the benefit of another person, family, region, or nation is all of one morality— "might makes right." Whether it is the use of military force by one nation against another nation, against a secessionist region, or to institute "ethnic cleansing," or the use of armed force by the tax collector to enforce death taxes against a family or income taxes against the individual to use the property for the benefit of others, it is all of that one morality— "might makes right."

Democratic socialism that tries to legitimize the confiscation of the property of one class for the benefit of another is a morality of slavery. As long as this is the morality, wars, armed suppressions, terrorism, rancor, hatred, and contempt will plague the societies of the planet. Only through free, open, honest marketplaces and volunteerism can truly free-orderly civilized societies evolve. A "might makes right" morality is incompatible with a free-orderly society. It fails to harm the person, property, and minor children of others, whether it is administered by a democratically installed government or by a despot. The "might makes right" morality has bestowed acrimony, strife, war, brutality, and genocide on humanity from the outset to the present. A democratic government is capable of exercising the "might" of a majority over a minority as defined by an election, and thereupon it is ordained "right."

A government that uses its coercive powers to confiscate and redistribute the production or property of individuals is a system of slavery—a system enslaving producers, at least in part, to the service of non-producers or lesser producers. In feudal societies, confiscation of a portion of the product of the producers created serfdom. Even a significant portion of the production of slaves was used for and by slaves. During the years of the republic, many Romans thought of their society as free, but there is little doubt that the producers, the slaves, did not hold to the same view. The same could be said of the democratic Southern states of the United States prior to the 1860s.

A government that threatens physical confinement in order to coerce the confiscation of property fails any test of respect for the person or respect for the sovereignty of its citizens. The collection of income taxes and death taxes are clear examples of government conduct that fails any test of respect for the person and property of its citizens. Indirect taxation, such as a consumption tax, can accommodate the few essential government functions that cannot be financed by "use taxes or fees." Government is just one of the many services that all citizens purchase. Government benefits all, and all must pay for it. Doing so in proportion to one's total purchase of goods and services would suggest a proportional benefit. There cannot be a morality for an individual citizen different from that of groups of citizens who control the government by whatever method. For a king to be exempt from taxation is the same as a majority in a democracy being exempt from any or all taxation. Here is a simple question in the form of three examples that might help one to grapple with the morality of an income or death tax:

First, a person breaks into your home with a gun and demands, "Give me your money."

Second, a person breaks into your home with a gun and demands, "Give me your money, I'm going to give it to my boss."

Third, a person breaks into your home with a gun and demands, "Give me your money, I'm an Internal Revenue Agent, and we are going to give it to the first person."

This begs the question that demands an answer: "Which one is moral?"

For those who are concerned with the disparity of wealth in a society, they must accept that the very nature of human beings naturally creates disparity. All have different abilities, talents, and initiatives. Of the nearly eight billion people on this Earth, we are all unique and unequal. Human beings must be accepted for what they are. Creating a social structure that fails to recognize the nature of human beings or wishing they were different would be like creating a shirt with three sleeves because it would be nice if humans had three arms.

Somehow, it has become a "given" in many societies that inequality of income and wealth is bad. Such thinking panders to the human motivation of jealousy. One must ask, "How did this, so base a human emotion, gain such lofty stature?"

The society that desires to reduce the disparity of wealth or income must not practice irresponsible procreation. Irresponsible procreation constantly produces an endless stream of impoverished children and impoverished families trying to deal with it.

The society that desires to reduce the disparity of wealth or income must teach and extol the virtues of saving. Wealth cannot be had or maintained through profligacy, and poverty will be the plight of children whose parents are not capable or willing to be responsible for them. This, which we have alluded to, is developed further in this writing.

Also, implicit in a free society, all religions must be practiced in a manner that respects the sovereignty of all persons over their person, property, and minor children, and the practice of all other religions that can be practiced in the same manner. Religions that fail this test are, by their nature, incompatible with a free society. Such religions might better be practiced in a society having a state religion. Within these broad behavioral requirements, many can adhere to other moral and virtuous behaviors. Within these broad requirements, virtually all religious tenets can be respected and practiced by all families. The many civilities and virtuous conduct toward others can be incorporated into and practiced by individuals within the code of moral conduct prescribed for those who choose to live in a free-orderly society. The tests for the approval or acceptance of any specific actions by individuals or groups of people are expressed as follows:

Moral behavior in a free-orderly society is that behavior which, if practiced in a like or similar manner by all citizens, would produce a free-orderly society.

All participants in society must do no harm to the person, property, and minor children of others. All who choose to live in a free-orderly society must accept and be governed by a morality that honors the sovereignty of each person over one's person, property, and minor children, and the practice of all religions that can be practiced in a like manner. Harm done to one sovereign individual by another must be reparated.

PART TWO
THE DURABLE GROWING ECONOMY

CHAPTER 3

THE PROBLEM

The evolution of the failing dominant economic structures has been a struggle among many conflicting concepts. Maximizing economic growth, preserving property rights, fostering free markets, government-sponsored special interests' efforts to erect impediments in markets, the confiscation of one person's production to give to another, unfettered expansion of credit, and the geometric expansion of the human population have created a toxic political, social, and economic brew. Intensifying the conflict was reliance on such opposing ideas as those of Karl Marx, Adam Smith, the Fabians, John Maynard Keynes, and the Austrian economists, among others. Stirred into this concoction was the universal political and economic acceptance of the hopelessly corrupting graduated personal income tax, estate or death tax, and business or corporation income tax by dysfunctional, tyrannous democratic governments and dehumanizing, regimenting, tyrannous pseudo-democratic governments.

Soon, when all previously known economic systems have failed, a new economic system is offered.

In just the twentieth century, societies experienced feudalism, communist socialism, mercantilism, and capitalism.

All have fatal flaws. A new economic system is offered in the subsequent pages that we will label the **Equilibrium Economy**. The failure of the world economy is a result of the fundamental requirement that it has been built on an ever-expanding bubble of credit. Developed here are concepts that can replace the failed credit-bubble economic structure with one that can provide a durable, ever-growing economy in near equilibrium that provides prosperity and wealth accumulation for all its citizens. A fundamental maxim of Equilibrium Economics is:

That which the land and capital produce—that which each earns—must flow as freely as possible to the consumers without impediments or anyone's judgment or discretion.

The share that the machines, computers, and robots (Productive Capital, PC) produce and, therefore, earn must flow, as seamlessly as possible, to the consumer. The present system transferring the share PC earns through redistribution via corporate income tax, highly graduated personal income tax, death tax, impediments in the labor markets, employer contributions to FICA Social Security, luxury tax, unsecured consumer loans, second mortgage loans, student loans, that portion of mortgage loans that exceed the underlying value, unsecured state and local government loans (General Obligations), federal deficit spending, and other fictitious assets, is untenable, inadequate, and corrupting. A simple, straightforward system is offered here. It requires major changes in corporate governance. It requires significant restraints on the use of credit by consumers, corporations, banks, and governments. It will change how society saves and invests. Over time, it will guarantee every citizen an income and financial security.

Prosperity, Freedom, and Order presents rational concepts for achieving a durable economic system that can function in near equilibrium, minimizing the disparity between the rich and those who will be the "not so rich." These concepts have the benefit of the experience of the failure of centuries of feudalism, the failure of a hundred years of experience with the many forms of socialism, and the failure of two cycles of capitalism over the past one hundred and fifty years. Certainly, we should be able to have learned something from these colossal failures. Many things have been tried that work, and from these we should also learn.

No society can be free without being orderly. Given the choice of chaos or tyranny, tyranny will always be chosen; therefore, we always use the hyphenated word *free-orderly.* The ever-present, eminent threat of tyranny that assures poverty for most and economic stagnation for all but a very, very few should be sufficient to cause.

The Gross Domestic Spending

Here we introduce a discussion that may help the reader cope with this view of the economy. We have become accustomed to the measurement of Gross Domestic Product, but properly named, it might be better called Gross Domestic Spending. In the United States, it seemed to concern no one that consumer spending comprised over seventy percent of the GDP measurement. How did consumption become more than twice the measurement of production? The answer is the perverse application of credit, along with the resulting inflation of asset prices that became skimmed to produce personal income. Unsecured credit created the illusion of wealth and fostered sumptuous spending. Inflated ephemeral asset prices enabled skimming while maintaining the illusion that the assets were still intact, or even greater, while producing great income streams that supported even more credit, higher asset prices, more spending, and so on. The fact that the world has accepted the dollar as a reserve currency has had no small role in enabling this profligate mindset.

The Credit Bubble

This world economy was built upon credit of all kinds. The most egregious has been unsecured credit that ultimately will not be repaid, for it is only repaid in the aggregate from the income created by more credit. This credit has been or will be repudiated in whole or in part. Systematic inflation provides a systematic repudiation of this debt.

This is offered to help some readers with this. Johnny starts school with two plastic cards—one credit and one debit. The debit card allows for a ten-dollar purchase, the cost of his lunch. He was given the credit card to use only in an emergency. On his way to school, he sees a THING that he has always wanted, for sale for ten dollars. Thereon, he decides to buy the THING using his lunch money debit card. Well, now it is lunchtime, and Johnny is hungry. Having no lunch money, he decides this to be an emergency and pays ten dollars for his lunch with his credit card. What a marvelous decision. He not only had a good lunch, but he also made the world nine dollars wealthier. There remains nothing. He ate them. And these are fictitious assets. But there remains another consequence: the demand and production of THINGS were thrown into disequilibrium.

This credit boom started in the United States in 1949 as the disequilibrium of the previous credit bubble (1860 to 1930) finally became cured. The domestic hardships that accompanied economic adjustment were completed, with minimal domestic rancor, coextensive with the hardships of WWII. At that time, the USA was the dominant industrial economy of the world. The forced forbearance of consumption during the war years left families debt-

free. It left state and local governments virtually debt-free. It left corporations virtually debt-free. The debt outstanding was that of the federal government. It, in large measure, was owed to the citizenry through zero-coupon (war) bonds. This became the foundation of the massive credit bubble economy that has lasted for nearly seventy years.
The massive credit has advanced the standard of living in the world well ahead of what can be achieved in a balanced economy in equilibrium. As joyous as it had been, it was, by its very nature, doomed to crash. This process was readily recognized when a family took on great debt that they could not repay, enjoyed driving a wonderful car, lived in a very commodious house, and ended up in bankruptcy. Such schemes became illegal after Ponzi's business model of the 1920s collapsed, along with Samuel Insull's public utility pyramid scheme. But it was ignored when businesses, banks, and governments of all levels collectively did the same, servicing outstanding debt with newly created debt. Unsecured credit is a flawed "building block" on which to build an economic structure. As it creates the illusion of wealth, it aggravates the sense of a wide disparity of wealth among the population; it accelerates the growth of an economy to unsustainable levels and eventually crumbles, bringing down the entire economic structure.

The Underlying Problem

The underlying problem is that the great, technologically advanced industrial economy can produce prodigious amounts of products. It is readily recognized that a portion of the production is earned by labor, and labor is remunerated for its portion, and thereby is consumed. But that portion of production that the machines-computers-robots earn must be distributed to consumers by some other method.

The basic problem is caused by the fact that the industrial complex, using extremely productive machines, computers, and efficient techniques, can produce vastly more than can be distributed to the population in a sustainable, balanced manner. And the portion that the machines-computers-robots earn that is retained by the few owners is beyond their consumption. This process of "forced saving" and thereby excessive investment causes leveraged capitalism to inevitably achieve what was known in the 1930s as overproduction and what is referred to today as excess capacity. Overproduction is a state where supply exceeds that which can be consumed in equilibrium. The investment in excessive productive capital accommodates inflated levels of consumption. This becomes further exaggerated by the obsession with "insufficient demand" theory, which spawns massive increases in unsecured credit and inflation of assets of all kinds. This produces a state of ever-increasing disequilibrium, which inevitably fails. The central banks' attempt to increase demand by lowering interest rates causes further excess investments in productive assets.

In the 1930s, there was a widely held suspicion that there was just "overproduction." Being somewhat counterintuitive, it was given little or no credence, and in the 1960s was dismissed with a deficient money supply explanation. This aided and abetted the "inadequate demand" view that has largely dominated economic thought for the past one hundred years.

No amount of monetary or fiscal (government) stimulus of demand will overcome the propensity of leveraged capitalism to achieve overproduction (vast excess capacity).

When the economic benefits of the work done by machines flow to a few people, how do we enable the population to consume the machines' output? At an earlier time, a person owned the tool, made the consumer product, and consumed it himself. Commonly, the labor, machine, and consumption were closely tied or at least not very remote. Well, we have accomplished the new distribution of prodigious production in a couple of ways.

Firstly, we use the coercive power of government to confiscate through taxes on that which the machine earns—corporate income taxes—and on the people who own these productive assets—graduated personal income taxes.

Thereon, these monies are doled out to people who consume the output of the machines-computers-robots. Unfortunately, this is a corrosively corrupting moral contradiction—using armed agents of the government to confiscate the income and assets of one citizen to be given to another. This was enabled by the Sixteenth Amendment to the Constitution.

Secondly, a colossal system of credit has evolved to further distribute the production of the machines among the consumers. This has taken many forms. The most destabilizing of those is credit that is not properly secured by tangible assets. Because its continuance depends on an ever-increasing income stream, there is no way to repay this credit except with the creation of more credit. This unsecured credit is a process of creating fictitious offsetting assets in order to enable production to exceed a level of equilibrium with distribution.

We find credit card debt, second mortgage credit lines, governments' debt—city, county, state, and federal—that flows to consumers, and student loans that are directly and indirectly consumed. Then there is unsecured corporate debt that, when combined with massive bank credit, inflates stock, securities, and real estate prices. This massive, ephemeral, illusory capital base, skimmed by capital gains taxes, death taxes, transaction taxes, commissions, and fees, becomes realized capital gains and is converted to an increasing personal income. This process is essential for the continuance of the unsecured credit expansion.

The mandating of the Federal Reserve to force employment to unsustainable levels, lowering interest rates to zero and below, further expands the disequilibrium in capital assets, their inflated pricing, and excess capacity. The inflation of these fictitious assets creates the illusion that, after being skimmed, they still appear to be intact, even though the underlying asset has not changed. The cities of New York, London, Frankfurt, Singapore, and others have thrived on the skim of inflated values of capital assets. These processes produced the income with which the unsecured debt is serviced.

These fictitious assets fill the gap between what the machines, robots, computers, and other capital assets earn and that which, by whatever other means, is finally returned to the consumers. These fictitious assets (wealth) keep accumulating as sovereign debt, state and local general obligation debt, unfunded pensions debt, unsecured corporate debt, and unsecured personal debt of all kinds. It cannot be serviced except by the creation of more such debt. In no way can such be deemed durable. To treat it as such is delusional. It will be repudiated, either by inflation or by a financial collapse. As this is written, the extreme measures of the central banks have created such vast excess capacity that inflation of consumer goods and services is not a possibility. Accommodated by unsustainable credit, the bloated industrial/commercial complex that was designed to produce humongous amounts of goods and services has put the world economy in a state of gross disequilibrium. Keep in mind that all these fictitious assets, regardless of their due date, are considered by the consumers as their "money." The vanishing of this enormous bubble of "fictitious assets" shall be so vastly greater than the government's ability to monetize debt, there will be deflation, and a financial collapse.

At this point, the reader will either be accepting our prognostication of the huge crash and collapse of the financial, social, and political structures or be reading this after the fact.

CHAPTER 4

SO, WHAT CAN REPLACE THIS FAILED SYSTEM?

A durable economic system will have many features that are in contrast to the preceding credit bubble system. Such changes will be enumerated in the proposed amendments to the Constitution provided in the last chapter. The new system must be the product of a Constitution, as no legislator-created law can overcome the forces that corrupt, pervert, or reverse them for personal gain. Greed and self-interest are vastly too powerful to deal with such matters through legislation.

Personal Estate Fixed Investment Trust Accounts

Saving and forbearance of consumption, while foreign to current thinking, are essential to a durable economy. Minimum personal savings are required. A workable amount, based upon the American seventy-plus years of experience with Social Security, would be around fifteen percent. It will produce personal financial competence that will enable the individual/consumer to participate in the ownership benefits of the industrial/commercial/business complex. Distribution of this production to consumers is an inescapable requirement of any concept of economics that is to approach equilibrium. Styled correctly, each successive generation will be more prosperous than its predecessor, and the need for Old Age and Disability Benefits will disappear.

All earned income shall be subject to a savings requirement of some fifteen percent that shall be held in a Personal Estate Fixed Investment Trust Account, the investments in which are directed by the owner. Once invested, funds in a Personal Estate Fixed Investment Trust Account may never be sold or liquidated. This will protect individuals from the temptation of immediate gratification, which is a deeply ingrained element of the human personality that prevents far too many people from advancing economically. Personal Estate Fixed Investment Trust Accounts shall never be pledged or attached and shall be impervious to any and all litigation. Personal Estate Fixed Investment Trust Accounts may only be transferred by will, probate, or gift to another Personal Estate Fixed Investment Trust Account or a similar trust for a not-for-profit organization. While this may seem counterintuitive, such an arrangement will avoid needless skimming in the form of commissions and fees, resulting from unproductive churning. The common illusion is that if a company is falling on hard times, the stock can just be sold back to the market. It is not. It is sold to someone else to bear the loss. Changing the ownership would serve no societal purpose. Beneficiaries of the Personal Estate Fixed Investment Trust Accounts are protected by diversification. The amount of benefits lost when investors fail to take advantage of exceptional long-term growth in the earnings of companies in diversified portfolios vastly exceeds the amount of losses resulting from the failures of mature investment-grade companies.

The required contribution to this account may be varied by plus or minus five percent, established by the Central Bank based upon its determination of the capital requirements of the economy. This will be reflected in dividend yields and interest rates. Investment of these funds in the Personal Estate Fixed Investment Trust Account may be in stock, not to exceed five percent of the total assets at the time of the initial investment. Fixed Investment Trusts, having diversification not exceeding five percent and never in derivatives other than other Fixed Investment Trusts, shall be available for Personal Estate Fixed Investment Trust Account investment, along with index mutual funds. Although these securities, once purchased for the portfolio of the trust, can never be sold or liquidated, they can be exchanged for Fixed Investment Trusts, within a family of funds, for others serving different investment objectives.

The portfolios of the Fixed Trusts offered to Personal Estate Trust Accounts shall be classified as having an investment objective of Very Aggressive, Moderately Aggressive, Neutral, Moderately Conservative, or Very Conservative. Personal Estate Trust Account assets may be transferred among these funds. Personal Estate Trust Account assets may only be transferred by bequest or gift to another Personal Estate Trust Account or to a similar trust account for a not-for-profit entity. They may never be sold or liquidated. This will serve to cause each subsequent generation to be financially better off than its predecessor. Assets in the Personal Estate Fixed Investment Trust

Accounts may be exchanged for ownership of Fixed Investment Trusts, and ownership of one Fixed Investment Trust or index fund may be exchanged for that of another. This provides flexibility while assuring asset accumulation as well as competitive pricing of management services for Fixed Investment Trusts and index funds.

These Fixed Investment Trusts may be open for additional contributions, but the assets, once acquired, may never be sold or liquidated. Dividends will always be paid or distributed directly to the owner, never paid to or reinvested by the Personal Estate Trust

Account or Fixed Investment Trust. This process will assure that the earnings of the capital assets will flow to consumers, an essential process for a durable economy in equilibrium. It will assure income for the disabled and aged, and will enable each successive generation to be more prosperous than its predecessor. Capital investment developed in this manner will be plentiful, without producing the excesses and vulnerabilities of credit finance expansion. Capital markets can achieve long-term equilibrium and be vastly more stable.

The forced accumulation of wealth across the entire population will be a massive leveling factor and will mitigate concerns about disparity in income and wealth. It will end the burdening of the young from providing for the old, avoiding the negative consequences of that. It will strengthen families financially and eventually replace the social safety nets.

Think of how much better it will be for the robots and other capital assets to support the elderly than, as they are today, supported by the workers who have the problem of supporting themselves and their children.

That which appeared in our original writing as "The Congressional District Fixed Investment Trust (CDFIT)" has been rewritten as follows:

The National Fixed Investment Trust

A Fixed Trust is one in which assets, once deposited, may never be liquidated or sold.

A federally chartered independent Fixed Trust* will be chartered. A shareholder account will be established for each Social Security number. All dividends received by the *NATIONAL FIXED INVESTMENT TRUST* shall be distributed regularly. The establishment of the *NATIONAL FIXED INVESTMENT TRUST* will assist in a more robust recovery of the economy of the new free-orderly society. For example, banks and corporations that are recapitalized with government funding will issue commensurate common stock to the NATIONAL FIXED INVESTMENT TRUST, with the ownership of the reorganized banks being held by the population rather than by some government agency or large conglomerated entities. The path to a sound economic structure will be served. Other enterprises or earning assets acquired or owned by the government, such as the TVA, may be capitalized with stock and distributed to the *NATIONAL FIXED INVESTMENT TRUST.*

Dividends received by individuals from the *NATIONAL FIXED INVESTMENT TRUST* will be deducted from all other personal government distributions. This will include Social Security and all other federal distributions to individuals.

No individual, group, or commission will have discretionary powers over the holdings of the Fixed Trust. It must be kept in mind that any power an individual has over another is subject to corruption.

Corporate Governance

A major upheaval in business practices and structure must accompany the change. For an economy to achieve a state where "that which is earned by the machines, computers, robots, land, and capital assets" is distributed to consumers directly, without convoluted schemes of redistribution and unsecured debt, the ownership of the business and its machines must be widely owned across society. To accomplish this, one element will be that corporate capital must come solely from common equity or stock of one class. This will accomplish several essential elements of the new durable economic structure.

Firstly, with corporations having only one class of capital—common stock—and tangible assets secured by self-liquidating debt, they will be financially stronger as well as have more stable earnings. Stock prices will be less volatile, and bankruptcy will be rare. Secondly, it will force business management to evaluate honestly the merit of additional capital. The organization of publicly held corporations or other legal persons must be such as to serve the best interests of a widely held stock ownership, as opposed to serving small groups of leveraged shareholders and/or managers. To deal with the issue of streaming the earnings of the machines to the machines' owners—the consumers—no less than ninety percent of the net income must be distributed as dividends. This will also provide the proper investment medium for the mandatory personal savings and investment requirements described further on. The minimum required payout of ninety percent must be provided for in the Constitution because it is such a large element that if its designation were in the purview of human beings—in this case, the Congress—it would further corrupt them. The designation of a percentage of earnings larger than ninety percent could be a legislative prerogative. The hoarding of corporate earnings and stock buybacks assures inadequate consumer demand.

Publicly held corporations (corporations of which the Board of Directors collectively owns less than fifty percent of the outstanding shares) may have no fewer than seven Board Directors, Trustees, or other governing members. The reason for this is to enable a diverse shareholder interest to be served while inhibiting domination by special interests.

A publicly held corporation's board members may not serve on any other unrelated governing boards or be officers of the corporation or any related corporation. This is to minimize conflicts of interest, cronyism, and the concentration of potentially corrupting power. Boards should create incentive compensations that produce the achievement of established goals. Incentives must be effective to avoid the bureaucrat syndrome. Incentive programs should continue well beyond retirement. Measurements of performance should be based on net earnings, not per-share earnings or share price. Share price measurements reward short horizons and instability. Deferred incentive compensation must be in the form of deposits to Personal Estate Fixed Investment Trust Accounts.

The charters of publicly held corporations must contain provisions of irrevocable preemptive rights. Preemptive rights were, at an earlier time, virtually universal. Special interests and managers, to enhance their power, have managed to do away with these. There are provisions in the charters that require new shares to be offered to the existing shareholders prior to being sold to the public or anyone else. This protects the shareholders' ability to maintain their positions in the company, avoiding having their interests diluted and preventing a myriad of other activities detrimental to the interests of public shareholders. Also, charters must provide for cumulative voting. This will enhance shareholders' ability to elect directors who will represent shareholders' interests.

Shareholders of publicly held corporations shall form shareholder unions to vote proxies for their best interests, but may never control more than twenty percent of the board of any one corporation. This will foster the constructive influence of the boards and enable shareholders to have competent representation. The custodians of the share interests of passive investors shall have no proxy vote. These provisions for corporate governance will correct most of the abuses of the corporate power structure. They enable the new institution of universal ownership of capital assets. They will minimize corruptive political influence.

Land

Long recognized are the three elements of production—land, labor, and capital. Also acknowledged is that the genesis of nearly all land ownership is a result of conquest of some sort—hardly a concept of "good title." This solution allows the "earnings" or economic rent of the land to flow directly to all of the consumers, an essential event if an economic system is to function close to equilibrium. This solution assures the highest use of the land over time, which is essential to economic growth. This will offer Native Americans the option to participate in every aspect of the new consumer-oriented economy. Here we are able to recognize the three elements of production, which are: the farmer's land; he and his family's labor; and the capital that includes improvements to the land, the barn, fencing, as well as livestock and equipment. With few exceptions, that which the farmer's land earns and that which the farmer's capital earns flow seamlessly to the consumer, the farmer's family. This is a significant economic relationship that this amendment codifies for all productive land.

All land will be designated "RESIDENTIAL" or "PRODUCTIVE." The title to all PRODUCTIVE lands will be held in a Public Land Trust, to be held as an investment of the National Fixed Investment Trust. All of the lands will be subject to fifty-year land leases. The existing productive landowners' title will be converted to a minimum fifty-year lease and a lien equal to the depreciated value of all improvements at the end of the lease. Initially, leases will be staggered through adding two-year increments by lot. This will cause the annual auction of leases to equal two percent of the total each year in fifty years. This arrangement protects the leaseholders. At the end of the lease, the present leaseholder will be entitled to bid an amount equal to the depreciated replacement value of the improvements. The present owner (or the lessee) holds a first lien in the amount of the depreciated replacement value of all the assets on the property. He can bid the amount of the lien or an amount to be the winning bidder, or receive a cash payment for the value he contributed to the property. Whatever amount of the auction price of the new lease that exceeds the first lien will become the rent paid to the Land Trust for distribution to the citizenry through the NFIT. The Land Trust that holds title to all of the PRODUCTIVE lands shall distribute, at regular intervals. This will cause all of the earnings of land to flow directly to the consumers.

This will avoid the disruptive accumulations of vast windfalls resulting from land appreciation as well as mitigate the vast disparities in wealth. In this way, all adult citizens, regardless of their wealth, will own and benefit from the ownership of all the productive land. In the case of the farmer, this would include such things as the house, the barn, fencing, the dam of the pond, the enriched soil, and any other improvements to the raw land. After the passage of fifty years, the farm is up for public auction.

Residential land will be titled to existing resident owners.

Native Americans shall be given the option to participate fully in all of the benefits of the amended Constitution by subscribing to these provisions through treaties.

Integrating concepts

Because nothing like this now exists, these concepts will strain the imagination of some. So, imagine that in 1936, instead of Social Security, the same monies were deposited into Personal Estate Trust Accounts (PEFITA) and invested in diversified portfolios of the ownership of major corporations throughout the economy. And that these corporations had one form of capital—common stock—that disbursed 90% of earnings to consumers and retained 10% for expansion. Imagine further, these PEFITA as being passed to the next generation intact, providing personal income for that generation, and so on. Now, virtually all persons would enjoy a reasonably stable income. For most, seeking employment would be optional, and wages that enticed people to work would be intrinsically fair and economically justifiable. With the exception of a "basic needs" safety net, government payments of all kinds would make no sense.

Under such a system, corporations must be governed for the benefit of the shareholders as consumers, not as managers, speculators, or accumulators of massive wealth. Keep in mind that if the capital required by new and growing companies had come entirely from the savings of the public, the benefits of the growth of large corporations would not be so concentrated in the hands of a few.

So, understanding these concepts, one can visualize the end of Social Security; the end of the multiplicity of welfare programs; the end of the need for massive enterprises to engage in skimming financial assets; and the mitigation of the amassing of wealth that causes the great disparity of wealth and income among the population. The decision-perverting, resource-wasting, politically corrupting income taxes would no longer be needed. It is clear to see that this system, which serves the consumer and avoids massive disequilibrium, must combine the concepts of consumer ownership with the reform of corporate governance.

Unsecured Credit

Unsecured credit shall be relegated to small loan companies that function with only equity. Unlike the present industrial and commercial complex that produces vastly more than an economy can consume in a balanced and sustainable manner, the economy that achieves durability must be an economy that approaches equilibrium. As such, it must function largely without credit that is not self-liquidating, secured by tangible assets. This would prohibit unsecured consumer credit, bank loans not adequately secured mostly by tangible assets, corporate debentures, and general obligation bonds issued by state and local governments, among others. By removing the ability of a few to commandeer great amounts of capital for their personal benefit that is not their own, the minimizing of leverage will largely end the extreme concentration of wealth. This would greatly reduce the disparity of wealth and income from the upper end. This will help mitigate that major social irritant. It is hard to imagine this being accomplished without enduring loud and lugubrious objections. The extreme at the impoverished end of the income and wealth spectrum is dealt with further on.

Economic growth must be a result of innovation, increased productivity, investment of savings, and greater available resources per capita. It cannot result from created credit that must one day be repudiated, one way or another. This new structure will also diminish inflation and the instability of capital markets. It will be stable, slower growing than the system which, with the steroidal stimulation of unsecured credit, creates a gross disequilibrium. Remember, in a credit-driven economy, inflation is a systematic repudiation of debts (a dishonorable silent theft), and the "boom and bust" cycle is inevitable.

Banks

A sound, reliable monetary and banking system is essential to a sound economy and a free, orderly society. A central quasi-independent bank can produce, provide, and preserve a reliable and stable currency, a durable banking system, and, along with the PEFITA, can control an adequate supply of capital. It should have no further mandate. The supply of money growth must be matched to grow no faster than the equilibrium growth rate of the economy in order to avoid inflation and other disequilibria. The debt-to-assets ratio (leverage ratio) of banks must be prescribed in the Constitution and be substantial enough to withstand any economic calamity. Those trusted with such potentially powerful authority must have no personal financial interest in the banks or other institutions over which they have control.

When making loans, banks are likely to be among the institutions that will be increasing the money supply. Banks would not engage in other businesses. The equity of banks must be substantial. A pure bank, having a large common equity, will require only minimal regulation. Well-capitalized banks, making only self-liquidating loans secured by adequate tangible assets, can provide a money supply and credit structure that is sound. History and experience have shown that when banks engage in other business activity, the attention of managers becomes distracted and their banking judgment tainted. Here again, the entrenched establishment is vehemently disinclined to yield its far-reaching powers over banking, finance, and other fields.

The adequacy of available capital for investment will be governed by the central bank's control of savings through the PEFITAs. If the dividend yield on PEFITAs rises to high levels, it indicates an inadequacy of the supply of capital, which would signal a need for a higher mandatory savings rate. If the yield of PEFITAs falls below historical norms, it indicates an excess of savings. The central bank will have the power to increase or decrease the withholding rate for PEFITAs to establish a reasonable cost of capital and a reasonable reward for foregoing personal consumption. It is here where we find the equilibrium of the supply of goods and services and the demand for them in our new economy.

Leverage

Corporations, including banks, other than investing in treasury securities, may borrow or lend only when secured by tangible assets and with not less than twenty percent equity. This provision would include deposits in banks.

Broker/dealers and other financial institutions, which shall be separated from banks, may have the prerogative to post intangible collateral. This will end the abusive and destabilizing practices of leveraging up the institutions, enabling the concentration of power, and weakening their capital structures.

Mortgage lending

A mortgage loan must not be renewable. It must be amortized and may not be increased when secured by the original asset. It may be refunded at the same or a lesser amount and must be transferable when approved by the lender. Improvements to a property may be funded by an additional mortgage lien that must be amortized. The financing of a real estate improvement must not be refinanced once paid off. The exception would be for individually owned residential property at the point of sale. This provision will avoid leverage abuses and advance the stability of equity ownership, thereby strengthening family finances. The mortgaging of a fifty-year-old commercial building is an unnecessary economic destabilization frequently precipitated by tax law. An eighty-year-old Empire State Building should be an equity asset and not loan collateral.

The Mutual Savings and Loan Associations

The Mutual Savings and Loan Associations can once again be a significant contributor to an Equilibrium Economy.

They finance home ownership with savings rather than creating money. As background, the Federally Chartered Mutual S&L was largely a creature of the early Great Depression years. It was a marvelous concept that thrived through the fifties, sixties, and seventies. Throughout much of the nation, S&Ls produced most of the home loan financing. They gave us the "thirty-year, fully amortized mortgage-backed loan." We now take this incredible innovation for granted. But prior to 1930, home mortgage loans accounted for barely one percent of the bank loan portfolio. Most home loans were private, five-year, balloon notes. This wonderful concept of financing homeownership with savings has now been virtually extinct for two generations. It may take some effort to resurrect it, but we believe it will be worth the effort.

They were local and regional, serving their communities not only financially but also by providing social leadership.

The demise of the Federally Chartered Mutual S&L was due to a combination of the Fed-induced inflation, its resulting devastating high interest rates, and the corrupt, perverse changes to expand their sphere. Aggressive expansion of eligible alternative investments combined with the attendant endless regulations produced unwieldy, unmanageable entities that ultimately collapsed. What we hope to do here is to present an outline of the original concept for federally chartered mutual S&Ls with modifications that will insulate them from the forces that caused their catastrophic collapse. This marvelous institution that finances housing with personal savings, structured correctly, will give the nation a sorely needed, strong, durable economic building block.

A totally new charter, separate from the old S&L structure, would be created. This will be totally separated from the convoluted structure and regulation of the past. It will be simple, restricted to home loans and U.S. Treasury and U.S. Treasury-backed notes to meet the liquidity requirement. The structure and regulations must be simple so they can be managed by people of average talent.

Here are some modifications that will make a healthy, durable Mutual Savings and Loan industry. These are based upon the fifty years of successful S&L experience from 1930 to 1980. They will not be stock companies. This will avoid the misincentivized management. Shares and share options held by management caused unsound practices and efforts to influence legislation that weakened the S&L structure. Their federally backed deposit insurance must be limited to federally chartered mutual S&Ls. State-chartered entities must not be included. As an insurer, this new entity will develop reserves that guarantee principal deposits. Appropriately, if the government agency, the Federal Reserve, creates inflation and a decline in the value of the underlying collateral assets—mortgages—the government can assume responsibility for the loss.

Mutual Savings and Loan Associations will never again have a specified dividend. The dividend payout will be limited to the net earnings. The promised "interest rate" payout became a promise that could not be kept. Commercial banks were limited to paying ¼% less than the S&Ls on CDs, furthering the misconception that the S&L payout was interest. The public will learn to accept the varying dividend payout as they create a history. The depositors are shareholders.

A mandate of the newly chartered S&L will be that all withdrawals will require at least a seven-day notice and may require as many as thirty days' notice. They will never again provide funds on demand. They are not banks and must not be perceived as banks. As originally conceived, nearly all of their deposits will be invested in first mortgage loans while holding minimal liquid assets. An "optional" thirty-day notice requirement for withdrawal appeared on all passbooks but was not mandated. This led to the general understanding that funds were available on demand. This led to checking and the liquidity disaster. S&L deposits, by their very nature, are investments in long-term loans and are not bank deposits. S&Ls will never again issue negotiable or transferable certificates of deposit with government-backed deposit insurance. In the eighties, using this structure, they were in effect issuing government bond equivalents. This precipitated unimaginable abuse.

S&Ls will never again make construction loans. These loans garner high interest rates because of their risk and a requirement for skillful, diligent administration. The high interest paid by the S&L for the construction loan came from its depositors. Paying themselves high rates of interest would have earned the envy of Charles Ponzi.

S&Ls will make no second or lesser lien loan unless subordinate only to loans held by it. These loans will be mostly home improvement loans. Appropriately, these will enhance the homeowner's equity and the S&L's collateral. The true merit of the S&L is the production of strong financial structures for retirees. Currently, refunding loans may increase the principal amount of the loan or lower the payment. Properly, they should only reduce the term of the loan. The societal reason for this is to strengthen households financially, not to encourage spending. Only home improvements should increase the indebtedness of a homeowner. Here, the S&L can guide the homeowners with prudence as they strive for financial security over their productive years. A home held free and clear is commonly the financial foundation of the average retiree.

To avoid the fate of our previous S&L experience, any future modification of their charters must be made difficult. It is for this reason that we offer it here as an amendment to the Constitution. Federally chartered Mutual Loan Associations must never be allowed to make political contributions.

A new S&L system shall be created by soliciting applications for charters from a specific number of contiguous zip codes in each congressional district by population. Selection of an application shall be based solely on the applicant's professional competence, experience, and community involvement. Political allegiance, race, gender, ethnicity, or age shall never be considered. The newly chartered Mutual S&L Associations can be allocated to each congressional district. The Federal National Mortgage Association and the Federal Home Loan Corporation can disassemble their mortgage-backed bonds and deposit them into the federally insured S&L covering the appropriate zip code. These loans, being on properties in their community or region, must be discounted to yield a rate that will entice personal savings and investment. In this way, a virtually unlimited amount of mortgage-backed bonds can be re-secured by S&L deposits. This will help facilitate the privatization of this government enterprise and reestablish a viable S&L system.

Income and Estate taxes

So, now let us deal with another dramatic change—Income Tax. The Sixteenth Amendment destroyed the sanctity of private property, a fundamental element of the freedom of the individual. Just as the vast bubble of unsecured credit was universally embraced, graduated income tax was also equally embraced. Somehow, the horribly destructive consequences of income taxes—that pervert nearly all business judgments, hopelessly corrupt government, and make prevaricators of the citizenry—were able to be ignored. The ostensible purposes, social equality and political advantage, will no longer be a problem in the new durable economy and free-orderly society that is disclosed here. There will be no need for the manifold destructive effects of income and estate taxes in order to redistribute the "machines'" share of the production to consumers or to mitigate the disparity of wealth and income. Graduated income tax and corporation income tax did the right thing, the wrong way, for the wrong reason.

The tax base that replaces it must be a soundly structured consumption tax. The government serves all citizens. As part of the cost of living in a society, the burden of paying for these essential services of government can only morally be borne by all citizens. Government services are a part of every citizen's consumption. Those who consume the most are presumed to enjoy the greatest benefit, so for them to bear proportionately more could be justified. Any such consumption tax must be levied on all consumption equally. Any discretion in how or on what services or products will be taxed will produce the rankest corruption. The benefits of any exemptions would produce such great monetary benefits to some as to be capable of corrupting almost any government official, and those who could not be corrupted would be replaced.

Income tax and estate or death tax make a mockery of any rational concept of property rights. Property rights to the fruits of one's production are essential to freedom. This is disclosed further on. As will unfold in this presentation, the massive social expenditure by the government will gradually disappear, and most worthwhile public capital projects will have non-tax financing, which will be self-liquidating. The costs of many necessary government-provided services will be borne by the beneficiaries of those services. Shifting the costs of government services has been just one more method of redistribution. The social and economic structure revealed here will diminish government involvement with the social "safety net" as the need for it gradually disappears.

Labor Unions

Monopolies granted to labor unions have, in the past, enabled the earnings of the capital assets to be distributed to labor/consumers. This is not necessary under the organization of the economy revealed here. But corporations have two classes of employees—management employees and operating employees—who have conflicting interests. For this reason, labor unions organized to the limit of one employer in one "market area" must be encouraged and protected. In this way, if they sink their boat, they all go down together. The labor union will not be able to sacrifice the jobs of one part of its membership for the benefit of another. However, this does not include, of course, government employees or public utility employees, or those of any other monopolistic employer.

Free, Transparent, Honest Markets

An economy must constantly adjust to changing conditions. Honest, transparent markets free of impediments are required to keep people free from the coercion of others and to achieve maximum growth.

Generally, such markets are contrary to human nature. People are inclined to erect impediments in markets that serve their shortsighted self-interest. For this reason, the government's production, protection, and preservation of honest, open, competitive markets is the only protection against such behavior. When the government fails at this task, little or no hope is left for a functioning economy free of corruption. So, even the government must be of such a style that it can withstand people's selfish nature. In this resolve, government officials, regulators, and bureaucrats must not have personal financial interests, directly or indirectly, that they can influence with their governmental authority and responsibilities. It is only in an open, honest, competitive market where self-serving people can come together and serve this very human nature. Regulation must be avoided where reasonable governance can be achieved by non-government entities, operating in a free, honest, open, transparent marketplace. As developed further on, these will include not-for-profit organizations.

Coercion in the marketplace is a product of monopolies and/or quasi-monopolies. To whatever extent it is within the power of government, government must not allow this. With an absence of coercion afforded by such a marketplace, the elements of production—land, labor, and capital—can be allocated in the most efficient manner. Without coercion beyond the establishment of the conditions of a free market, the marketplace can distribute the products among the population. The government must be involved to free the market of fraud, deception, and non-competitive tendencies and practices. Matters of natural monopolies must be within the purview of government in order to avoid the coercive and abusive practices that they engender.

It has generally been deemed essential to economic growth and development that invention and creativity be rewarded with limited property rights, granted by government in the form of patents, copyrights, and trademarks. These do create limited monopolies in the marketplace, without which creativity and inventiveness would be stifled. Here again, the periods for the granted monopolies and the extent of those monopolies—patents and copyrights—must be limited to that which is necessary to serve as an incentive for creative firms or individuals. Free people can only interface in either a free, honest, informed marketplace or as benefactors and beneficiaries.

Foreign Trade

When foreign trade is for productive capital assets, it serves the future growth of an economy. But when foreign trade is at a deficit, acquiring consumer goods and services, the transaction is a trade of claims against capital assets for consumer goods. When a society is consuming its wealth, on its face, it is not a state of a durable or balanced economy. Here, a solution is fairly simple. Government, after adjusting for net currency reserves, needs only to raise the tariff on all consumer imports at regular intervals, without discrimination, until the trade balance of consumer goods is zero. The tariff should be reduced to zero as the process becomes reversed. Trade in productive assets is not considered in the consideration of a negative trade balance, as they are assumed to be self-liquidating. If any authority for discretion is granted, it would inevitably corrupt the process and all of those who are involved. In a society where the tax base is a consumption tax, the tax needs only to be reduced proportionately to offset such a tariff. Tariffs that deal with aggressive and unfair trade are a distinctly separate matter.

Retirement

The institutions of age-specific retirement, either public or private, will not be part of a durable economy in a Free-orderly society. Retirement will have to be a decision made by each individual, as he/she and the employer evaluate numerous considerations. And keep in mind, longevity is increasing. To what age, we do not know. This may be of little significance as the required savings in Personal Estate Fixed Investment Trust Accounts accumulate.

Institutionalized pensions will not be part of such an economy, as they are commonly contractual arrangements made by and for one generation to be paid by the next. This is untenable because it burdens or enslaves the subsequent generation into the service of its predecessor. As self-reliance is fundamental to freedom, saving, insurance, and annuity contracts will be the foundation of retirement and disability protection.

In a society where the population is not increasing, such as currently in Japan, Europe, and China, the labor force will become older. But keep in mind, the machines, computers, and robots are doing an ever-greater amount of the work, and the consumer population owns an ever-greater part of that production. So, as computers, robots, and machines do ever more work, people will do less. But because people are mandated to own that production and the earnings must be paid out in dividends, production and consumption can achieve a sustainable balance.

Don't jump to conclusions here. The Personal Estate Fixed Investment Trust Accounts are not the entire capital base. There will be normal markets for vast assets not held in the Personal Estate Fixed Investment Trust Accounts.

Population

This brings us, as we promised, to these earth-shaking innovations. They are a very bitter pill for the entrenched establishments. However, there is a point where it is intolerable not to accept change regarding population and the matter of "Irresponsible Procreation" in a durable equilibrium economy. No society can be free, orderly, or enjoy economic durability while tolerating Irresponsible Procreation. It is basic that the human infant cannot care for itself. Therefore, a society that is unwilling to let infants die uncared for must have a system assuring that each infant is cared for. To force men, who have elected to be sterilized, to restrict the size of their families, to support the progeny of those who cannot or will not support their own, fails any sense of a rational, free-orderly society.

No such system can be rationally described as a free society. To maintain a free-orderly society, society must require that each newborn go home with two or more competent adults, who have entered into a mutual agreement to care for the child until he/she attain adulthood and who are deemed reasonably capable of fulfilling that commitment. No societal structural change could do more for the well-being of children. We have more to say about the implementation and enforcement of this policy in subsequent chapters. Ending Irresponsible Procreation will greatly reduce poverty and thus the disparity in wealth and income.

Another question that we might deal with here is: "If everyone has income, how does the economy produce people willing to work?" The answer is that the market will take care of that. People who are inclined to improve their situation will be available to take jobs. The demand of the market will produce a price that will produce the supply. These employees will not be underpaid and will not likely be overpaid for long. The income level established for Responsible Procreation will likely be influenced by the level at which people with ancillary income will take jobs, as well as those with a genetic inclination to do so.

Now we have the basic "bookends" for the disparity of wealth and income. The bubble of unsecured credit creates the massive wealth at the top, while Irresponsible Procreation generates the poverty at the bottom. Consumption of personal wealth leaves the middle class feeling deprived. The provision above for accumulating wealth will continuously close the gap.

If a society is worthy of continuing, procreation is the highest work of life. If a society is to achieve order and prosperity in a state of continuance, it must maintain an adequate population while preventing unbridled growth in its population. We have more to say about this further on.

Here is an immutable law of Equilibrium Economics:

That which the land and capital earn must flow as seamlessly as possible to those who consume their production.

Inserted here is work completed subsequent to this publication.

The PERCEPTIONISM® Healthcare Plan

This Plan accommodates employer-provided healthcare insurance.

- This healthcare plan provides everyone in the United States with the same access to the best healthcare in the world.

- Everybody pays for healthcare with equal dollars. By providing the funding through a Value Added Tax (VAT), the person who consumes $300,000 a year makes the same relative payment as the person who consumes only $30,000 a year.
- This plan is not a single-payer plan. It is a 200 million plus 1 payer plan. Every consumer of healthcare will pay 10% to 20% out of pocket (cannot be insured away) at the time of usage, forcing market discipline on the healthcare system.
- This plan will enhance market discipline by requiring transparency—the honest disclosure of pricing.
- This system will force market discipline on the pharmaceutical industry and those healthcare providers that have a monopoly in their market.
- This plan provides for an orderly transition, relieving employers from the burden of responsibility for healthcare.
- This plan will free the government of healthcare responsibility and get it out of the corrupt healthcare business.
- This healthcare plan will ensure that everyone will have a Healthcare Savings Account available for healthcare expenditures for their old age when their need is greatest. It increases the sensitivity of the elderly to healthcare costs, enhancing market disciplines.

SELF-MANAGED MEDICARE FOR EVERYONE

A COMPREHENSIVE HEALTHCARE SYSTEM

What rationale makes some eligible for this total care system at the expense of others, and not for people under the age of 65? This is not rational. It is the product of a convoluted system of SPECIAL INTERESTS that has corrupted our government.

Here is a fact that everyone must recognize: all goods and services are "scarce." That means they can never be free like the air we breathe. They will always be rationed in one of three ways:

1. By price and market forces.
2. By availability—queue and stand in line.
3. By quality.

Under this program, an independent Federal Healthcare Trust will be established to hold and manage funds designated for healthcare.

This program enables affordable market disciplines while avoiding queuing and lowering the quality of healthcare for individuals. Medicare will be available to all residents of the United States. Medicare will pay 80% to 90% of all healthcare costs. Ten percent to twenty percent (depending on one's relative income in each market area) must be paid out-of-pocket at the time and place of the purchase of a healthcare product or service. This cash co-pay percentage shall be adjusted incrementally until the total consumption of healthcare equals a percentage of total personal expenditure established by Congress. This cash co-pay cannot be waived. All vendors or service providers of healthcare must publish and post the lowest price at which they sell or provide their services. This must include the prices of all prescribed medicines, tests, appliances, or equipment. Penalties for the violation of selling that which is prescribed at a price below the published and posted price must be punitive enough to prevent violation. Healthcare providers must have available lists of those charitable institutions that offer aid in their field to special needs.

All employers that provide healthcare to employees will pay an amount equal to their healthcare expenses immediately preceding this program to the independent Federal Healthcare Trust Fund, and this will be diminished by 10% of that original amount for each of the following ten years. This will allow for the gradual adjustment of both the corporations' finances and the healthcare budget during this ten-year transition period. Because there will be no healthcare insurance, insurance companies providing healthcare insurance may obtain reimbursement from the Federal Healthcare Trust Fund for their operating costs resulting from their discontinuance of operation in their healthcare insurance businesses.

Drug companies that provide drugs subject to patents must be totally separate entities from companies that provide generic drugs. This will help avoid predatory pricing. All drug providers must sell no drugs in the USA at a higher price than the lowest price at which it is sold at any place in the world. This way, all countries pay for the development of the drugs, not just the US population.

Healthcare providers that have a monopoly in any market area shall be subject to a Federal Price Control Board. This Board will establish prices based on prices in "price-disciplined" markets, adjusted for differing costs and costs of living.

So, how is the 80% to 90% paid for? Well, everybody consumes healthcare, so everybody must pay for healthcare. It becomes part of everyone's consumption as it is paid for by a Value Added Tax (sales tax), a consumer tax on all consumer goods and services. Some will pay much more than the value of the healthcare they consume because they are blessed with good health, or they are blessed with greater income and therefore greater consumption. The VAT tax rate shall be adjusted at regular intervals to accommodate the requirements of the Healthcare Trust Fund.

No exemption from paying the VAT/sales tax shall be granted, as any exemption to participating in the VAT/sales tax will open the door to the rankest corruption of government.

For Employers providing Insurance

Recognizing that the entire Medicare portion of healthcare costs is funded by the V.A.T., payments to the Trust Fund, the healthcare provided by the employer can be reimbursed by the Trust Fund. As we know that the nation's healthcare is x% of personal consumption and we know what the personal savings rate is, we need only to deduct the personal savings percent from the entire payroll of the insured company and reimburse the company the same percentage of the total healthcare costs paid by the Trust Fund. As long as insurance companies' products serve employees better than the "Self-Managed Medicare," insurance companies will thrive.

Federal and State contributions to Medicaid will be drawn down to zero over a period not exceeding ten years.

Increasing the effectiveness of market forces is important to a successful program. This can be achieved by raising the vested interest of each consumer in their own healthcare purchases. This can be achieved as follows: The 1.46% now collected for Medicare as a payroll tax will become instead 1% of the VAT deposited in a Personal Healthcare Savings Account - Trust (PHSA-T) for each individual. These PHSA-T funds shall not be used for any purpose and will not be accessible for healthcare expenses until age sixty. The individual's PHSA-T, after age sixty, shall vest and only be disbursed for healthcare expenditures or unencumbered to testamentary beneficiaries.

Commencing at age sixty-one, the 10% to 20% out-of-pocket cost to the individual will increase 0.5% each year thereafter. This process will place an ever-greater "market force" on healthcare costs to an individual, to a point where the welfare of one's heirs is greater in value than the cost of a specific healthcare service to that individual. As part of the implementation period, this 0.5% annual increase to the out-of-pocket provision will be waived for persons over forty years of age at the time of enactment, as well as for persons currently receiving Medicare benefits.

This is not a single-payer. It is a 200 million Payer.

CHAPTER 5

THE ROLE OF GOVERNMENT IN A PROSPEROUS FREE-ORDERLY SOCIETY

At the beginning of the twenty-first century, there exists a vague, muddled, conflicted understanding of what constitutes individual freedom or a free society. These notions of freedom rest upon a comparison of democratic societies with the tyrannous monarchies and dictatorships still in existence or fresh in the memories of those who witnessed the twentieth century directly or through the descriptions of living ancestors. While functioning under a belief that they are free societies and without any clear understanding of what individual freedom and a free society are, the democracies have moved on a relentless path to self-imposed tyranny. Democratic governments, driven by a variety of motivations, have imposed their coercive powers on virtually every aspect of human activity. One needs only to extrapolate this process to foresee an intolerable tyranny. That which follows will challenge the reader with a significantly different understanding of government when viewed from this conventional notion of freedom and the prevailing understanding of "good" government.

Because the government provides the foundation of a Free-orderly society and at the same time is the greatest threat to individual freedom, the presentation in this chapter is the most voluminous.

The scope of government presented here deals with the highest level of government – generally national government. Lower levels of government will deal with provincial and local matters that are generally unique to the state, province, region, locality, or community. And they will do so within the limitations placed upon the supreme government and within the morality of a Free-orderly society. *The highest level of government must be restricted to only those matters essential to produce, provide, perfect, protect, and preserve a Free-orderly society.* Any expansion beyond these requirements invites and assures greater corruption and encroachment upon, and the destruction of, individual freedom. These obligations of the national or supreme governing bodies must not be duplicated or mitigated by any local or regional governments.

It cannot be stressed too strongly that inasmuch as freedom is a state-of-being, government exists solely to enable, produce, provide, perfect, preserve, protect, and defend that state-of-being—the freedom of the individual in a Free-orderly society. It is not and cannot be the role of government in a Free-orderly society to serve some real or imagined "general welfare." Freedom for individuals and the "general welfare," beyond that stipulated above, are in natural conflict, and government cannot successfully serve these two masters. To serve one is to deny the other; therefore, a Free-orderly society must deny responsibility for the "care and compassion" of its population. The grant of freedom places the responsibility for one's welfare and matters of care and compassion on each individual, the family, their religious group, or other voluntary associations committed to serving the welfare of individuals or families.

Societies having no other direction or mission might naturally be expected to embrace love, caring, and compassion for others as the mission of their society. While embracing love, caring, and compassion is a fulfilling endeavor meeting the needs of individuals, other individuals cannot be conscripted to serve such causes in a Free-orderly society. As has been developed previously, each person in a Free-orderly society must have command of his/her own resources of time and money to allocate among his/her own needs in pursuit of his/her own happiness. If providing care and compassion for others fulfills their individual needs, then they should choose to do so.

A system that might be properly labeled "Humanitarian Socialism" has become the moral foundation of many who have no other abiding religious convictions. As with any religion in a Free-orderly society, it must be practiced in a manner that respects the sovereignty of each individual over their person, property, and minor children, and the practice of all other religions that can be practiced in a like manner. Implementation of the care and compassion espoused by a Humanitarian Socialist government is by its very nature anti-freedom. It involves using the coercive powers of government to benefit some at the expense of others.

The needs for care and compassion of a society are limitless. It is a task beyond the capacity of any government. Any government that undertakes the task of providing care and compassion for its population will fail.

Understanding that freedom is the absence of coercion, one can compare heating and cooling to expanding or contracting freedom. To make a space cooler, heat is removed. To make a place freer, coercion is removed. To make a space hotter, heat is introduced. To make a place less free or more tyrannical, coercion is introduced. In the twenty-first-century democracies, there are thousands of lawmakers and bureaucrats who work tirelessly making laws and writing regulations, and thereby introducing an endless stream of coercion into society. If continued and propagated, this practice will cause ever-decreasing freedom in those societies, to the point of total democratic tyranny. It is estimated that there are about 4,500 criminal statutes in the United States and another 300,000 regulations that carry essentially criminal penalties, including prison. This comes about as a result of the government assuming responsibility for matters of care and compassion, assuming the role of guardian of the population, and, of course, the selfish interests of those who govern.

All who serve at the highest level of government of a Free-orderly society must do so motivated by virtue and respect, not for personal financial gain. It is therefore implicit that they have achieved their personal financial goals. Those who serve do so as "guardians and trustees of the freedom" for all the individuals in the society. To do so is to serve all the citizens of a Free-orderly society. To have any other purpose is to serve one group at the expense of others—an anti-freedom endeavor. Such a government, dedicated to a Free-orderly society, serves all who are so dedicated without strife, acrimony, and rancor. All who oppose such government, committed to a Free-orderly society, are implicitly and manifestly corrupters seeking special privileges at the expense of individual freedom and the well-being of others. Government leaders collectively committed to the development and preservation of a Free-orderly society do not constitute a political party. This is because they neither seek nor pursue any personal advantage at the expense of others in society.

Inasmuch as the fact that the government of a Free-orderly society cannot engage in matters of care and compassion, and these matters must be delegated to a multitude of private providers commonly serving specific needs, those persons guiding government must have a high quotient for being personally guided by reason, while persons guiding institutions providing care and compassion must have a high quotient for being personally guided by compassion.

Governments have found direction, guidance, and legitimacy in religion. Government separated from religion, as in "separation of church and state," leaves a state with no moral foundation and thereby no moral structure other than that which has been pointed out as "might makes right." In a Free-orderly society where there is to be freedom of religion, the questions arise: "How can government find direction where the values of one religion are in conflict with another?" "Which religious values are going to govern?" The concepts presented here of a Free-orderly society provide a common denominator for a moral foundation, a universal moral structure, and a moral compass for government that enables the practice of virtually all religions while avoiding the imposition of the practices of one religion upon the practitioners of another. If a political party were to become a religion and come to power, one could only expect the rankest tyranny and destruction of individual freedom. Nazi Germany comes to mind. In light of the preeminence of a Free-orderly society, the following will explain how government can achieve this end.

This moral foundation, structure, and compass of a Free-orderly society are found in the nature or characteristics of what has been historically labeled "natural law." Natural law consists of the most fundamental, self-evident, universal truths by which a society can or even must be governed. It is the bedrock, unyielding standard that can be administered universally. It is so basic and fundamental that it cannot be applied differently to the population. Here, the terms "axiom" or "axiomatic" will be used to refer to fundamental, self-evident, universal truths by which a society that chooses to be free and orderly must be governed.

- It is axiomatic that freedom precludes the coercion of one individual by another or a group of individuals by others.
- It is an axiom that freedom cannot exist without social order.
- It is an axiom of a Free-orderly society that while life is very precious, wherever and whenever life is in conflict with a Free-orderly society, life must yield. For there exists overwhelming evidence from human experience that establishes the fact that when and where the preeminence of individual freedom is compromised for the preservation of life, the erosion of individual freedom is limitless.
- It is an axiom that all persons in a Free-orderly society may do, say, or believe whatever they choose, but they must do so in a manner that does no direct physical harm to the person, property, or minor children of any others.
- It is an axiom that in a Free-orderly society, the freedom to exercise one's authority and the responsibility attendant thereto by each individual must be coextensive.
- It is an axiom that in a Free-orderly society, religious law may be practiced as long as it can be practiced in a manner that honors the sovereignty of each individual over one's person, property, and minor children and does no harm to the person, property, and minor children or impairs the religious practices of any others.
- It is axiomatic, therefore, that to achieve a Free-orderly society, sovereign individuals must give up to government that sovereignty that is required to preserve the remainder of their sovereignty.
- It is axiomatic that free individuals living in a free society will be a part of a society that maximizes the achievements

of humanity, and all of human experience confirms this axiom, and no human experience denies it.

Upon these axioms, all laws of a Free-orderly society must rest. No laws of a Free-orderly society, however established, can abrogate, disregard, or deny these axioms.

Today's technology for identifying and monitoring each individual assures societies will either be free in the style disclosed here or, most likely, total tyrannous societies. A "chip" placed in every baby will enable the reliable identity of every person as well as his/her whereabouts. In a Free-orderly society, such technology will produce phenomenal efficiencies, while in a tyrannous society, it will produce unimaginable repression and enslavement of the population.

Sovereignty of the Individual

The sovereign king or dictator has total life and death authority over his subjects and control of all of the property in his domain. In a free society, sovereignty must be vested in the individual. Any compromise to the sovereignty of an individual is a submission to the coercive power of another.

It is also a natural law that in a Free-orderly society, each adult must be sovereign over his/her person (body), property (production), and minor children. This sovereignty must be absolute and compromised, as mentioned previously, only to the extent necessary to preserve it and to enable sovereign individuals to interact socially and economically in an orderly manner. It is for this reason that minor children, incapable of assuming the authority and responsibilities of the sovereignty of individuals, can have no stature under the law. The only law governing minor children in a Free-orderly society must be the law established and administered by the parents. Within the family, in a Free-orderly society, authority and responsibility may be shared in any manner the participants may choose. Governmental intrusion into the governance of minor children becomes an open-ended usurpation of parental authority and sovereignty. The Western democracies have demonstrated the near-total destruction of parental authority by the intrusion of the state and thereby destroying the freedom of parents beyond any semblance of that which could be defined as a free society. Governmental intrusion into the governance of minor children inevitably intercedes in the practice of the religion of the family. Intrusion by the state into matters regarding the children usurps the authority and sovereignty of the parents, as well as shifts the responsibility through taxes to others, intruding on their right to sovereignty over their own property. No two families govern their children in the same manner. Therefore, in a Free-orderly society, outside of a religious group, none can stand in judgment of others, regardless of how inept the governance of some minor children might seem. It can be expected that within a religious group, the values of that group will be asserted upon its members as they govern their minor children. It is the inclination to value life over freedom that induces people to support governmental intrusion into family governance of children. Remember that in a free society, freedom must take precedence over life.

The government of a Free-orderly society exists as a result of sovereign individuals granting power to the government and, as a result, relinquishing some of their sovereignty. This is done only to the extent necessary to preserve the remainder of their sovereignty. All who benefit from government must relinquish sovereignty on an equal scale if government is not to be coercing some citizens for the benefit of others. Only those relinquishing their sovereignty can be involved in the determination to do so. That is to say that only sovereign persons paying a tax or submitting to a regulation may vote on such measures, either directly through a referendum or indirectly through their representatives.

All authority (rights) is vested in the sovereign population except for those specifically granted to the government by sovereign individuals. The sovereign population must specify who will administer the authority granted to it. This is important on the local governmental level, as local governments will be involved with matters peculiar to that area. National government, being restricted and dedicated to matters preserving the sovereignty of the individual, can best be served by persons selected by local governments. Individual citizens, tending to their personal responsibilities, are less qualified to evaluate and select representatives who will perform this function. This is the foundation of representative government. And this will include the administration of how the accused must be treated, the limits upon the authority of government regarding search and seizure, habeas corpus, and any other specific encroachment upon sovereign individuals that are not specified in the Constitution. A government that respects the sovereignty of each citizen must engage with each citizen equally. It is by this concept that "all men (and women) are created equal."

Individual sovereignty over one's person and property is the foundation of freedom, without which freedom of the individual cannot exist. The compromising of individual sovereignty during the twentieth century in the democratic states became the open-ended erosion of individual freedom. It became an abominable bureaucratic tyranny disguised as legitimate by the mantle of democracy. A reminder to the reader: If you are not sovereign over your person, property, and minor children, someone else is, and they are your master—and most likely, you are their slave in some manner.

Corruption of Government

Totalitarian government is extremely subject to corruption. Democratic government, while less exposed to extreme corruption, is nevertheless exposed to vast corruption. There is nothing in democratic government that protects one group from being taxed or exploited for the benefit of another. The beneficiaries can, through organizing numbers of voters among their group or through the influence of the media, organize voters outside their group to cause the government to serve their own ends at the expense of others. By any reasonable standard in a Free-orderly society, for government to serve one group at the expense of another is corruption. More directly, corruption of government is achieved by influencing the political representatives or the bureaucrats. It fails the requirement of a Free-orderly society to not harm the person and property of others. This is no revelation. It has been a part of human knowledge and discussed widely since the fifth century BCE. *The only protection society has from corruption in government is through minimizing the role of government itself.* For whatever power is vested in government will be converted, to some extent, to serve the well-being of the politicians, bureaucrats, and governors who wield it. As the government expands its activities, it broadens its coercive powers over the citizenry, diminishing freedom. In a Free-orderly society, government activities, therefore, must be limited. There are, however, a few responsibilities of government that are essential to a Free-orderly society. Therefore, the supreme government must be limited to only those functions that are essential for a Free-orderly society. Safeguards against the abuse of even limited powers must be imposed. A rigid Constitution imbued with an unwavering commitment to the axiomatic principles of a Free-orderly society, along with the separation of governmental powers, is apparently the only structure of government that is capable of creating and preserving a Free-orderly society.

Almost universally, those who would support a socialist government find that when it fails, the problem is a result of those running the government, never the system. The fact is that human nature and socialism are incompatible.

Socialism is a system designed for human beings who do not exist. As it has been pointed out, to tailor a shirt with three sleeves because it would be nice if human beings had three arms is the same as styling a socialist system for a human society. Human beings will always be governed by the pursuit of their personal Needs and desires. Persons whose temporal and psychological Needs are sated could serve without corruption, but by their very sated nature, finding those willing to undertake government service is extremely rare.

In the twenty-first century, Western democracies are fatally flawed as a result of their attempts to provide care and compassion for some citizens at the expense of others. The task of providing care and compassion is without limit, and the recipient groups have become so large or powerful that politicians are unable to resist the ever-increasing demands by an ever-increasing number of voters. The politicians solicit votes by promising, "If you vote for me, I'll give you this, and I will give you that." The result of this corruption is an ever-diminishing of freedom, ever-declining productivity, and economic stagnation. The more responsibility that individuals shift to government (their fellow citizens), the more authority government will exercise over their lives as it tries to extricate itself from the quagmire.

The founders of the United States of America had not contemplated, or at least not provided for, the deleterious consequences of political parties. *Political parties are coalitions of conspirators who use the coercive power of government to their own advantage or to serve their agenda at the expense of others.* The very nature of such coercive activities is anti-freedom, serving to corrupt and destroy a free society.

The term "civil war," as the words are used in contemporary language, is quite an accurate description of our modern-day politics. It is truly a civil power struggle, conducted quite civilly, where each side is trying to overpower the other for their collective personal gain. The stakes are commonly so large that it is little wonder that elections deteriorate into violence in many fledgling democracies.

Discretionary income taxes in the hands of professional politicians have hopelessly corrupted the Western democracies. One only needs to extrapolate the process of the daily promulgation of laws and regulations that govern every imaginable facet of the lives of citizens to recognize that, in some extended period of time, there will be no recognizable individual freedom.

The belief that government has a role in providing "care and compassion" causes the government to become the guardian of the people. This has spawned all manner of governmental bureaus, agencies, departments, and so forth. This becomes the most rancid culture for corruption. The only way to avoid this governmental cancer is to avoid its existence of such. The only bureaus, agencies, and departments must be those that serve the essential functions of government as it provides, preserves, protects, and defends the sovereignty of its population. Here, great care should be taken and procedures put in place to minimize the inevitable corruption.

Essential functions of government

Essential to a Free-orderly society is an economy that is sustainable in a state of near equilibrium. It must be one in which all of the citizens/consumers own and receive the benefits - earnings of the productive "machinery" robots and other capital assets - and, not an economy that relies on the endless creation of fictitious assets, to distribute its production. *Because of the preeminence of individual ownership of the productive capital assets requires saving/self-denial and that this ability is not universal among the population, personal sovereignty of one's property must be compromised by a requirement for saving and accumulation of wealth.*

Essential to a Free-orderly society is the requirement that all people function in a manner that does not harm the person, property and minor children of all other people. Government involvement in assuring such behavior is one of its essential and legitimate functions. These basic standards must be universally consistent throughout the society and therefore must be established and administered at the highest level of government. Inferior governments can deal with less basic matters that are peculiar to their jurisdiction.

Essential to a Free-orderly society is the formation of strong, competent, autonomous families. The family is the basic building block and foundation of a Free-orderly society. It is beyond challenge that the family is the most fundamental economic, social, educational and child-rearing institution of any stable society. The design and structure of the family must remain in the purview of the participants. Government must play a role in the formation only of strong families capable of fulfilling their responsibility of rearing the next generation of a Free-orderly society. *Healthcare being the domain of the family, if deemed essential by a society, a system of financing healthcare that is otherwise consistent with these principals is offer by The Institute of Perceptionism, Inc.*

Essential to a Free-orderly society are the establishment of private property rights. People must have total authority and control of their own work product. Anything short of that is some form of slavery or serfdom. The establishment of the structures of free, competitive, transparent and honest markets and commerce are the essential institution for free people to efficiently exchange the own work product. Other than Volunteerism, it is the only way free-sovereign individuals can interface with equality. Government must assure the ability of persons to enter into any contract that does not harm the person and property of others and for government to assure its enforceability. This may also accommodate contracts that provide for a private adjudication in the event of non-compliance. Adjudication must be timely, as history has shown conclusively that lawyers are ingenious in delaying litigation and thereby avoiding adjudication for their clients. Much of these matters have already been covered.

Essential to a Free-orderly society is its defense from external or internal threat. Not only may external forces threaten a Free-orderly society, but also there will be those who wish to take power for their own interests or cause and will attempt to seize power from within and thereby terminate a Free-orderly society. Also, defense against extreme disaster that threatens a Free-orderly society such as the earth being impacted by a devastating meteorite, biological epidemic, geological or metrological catastrophe must come within the purview of government. *As demonstrated throughout history conscription may be an over-riding requirement for the fulfillment of this provision.*

Essential to a Free-orderly society is the conduct of foreign relations. The government of a Free-orderly society must exist and function in a world of tyrannous governments both democratic and totalitarian. Guided by a might-makes-right morality, each will view a Free-orderly society as a threat either from holding the high moral ground or as a competitor in the world markets. The purview of a government of a free-orderly society does not reach into the governance of other societies.

Essential to a Free-orderly society is the ability to designate the preservation of forests or other lands to be held in perpetuity as "preserve lands". This would include the preservation of the many life forms that reside thereon. Also, essential government buildings and lands as well as parks and monuments can be held and preserved in perpetuity by government.

These are essential functions of the supreme government in a Free-orderly society. In order to minimize corruption and encroachment upon freedom, all of society's other functions must be relegated to the markets, private institutions, families, religious organizations, charitable organizations and regional or local governments

Practical application by the government to achieve these ends

Government shall enable and foster Rating Services to replace the free market-inhibiting business and professional licensing, admission-limiting associations, and accreditations. Government shall not engage in activities that can be relegated to non-government entities. For example, beyond establishing the rules of the road and the physical characteristics of the highways, with the electronic ability to toll highway use, the interstate system and its feeders can be divided into rational market segments and be placed in non-profit organizations governed by representatives of the cities, counties, and major business entities contiguous to and having a local vested interest in the tolls and capacity of that segment. This could disburse control of all the highways to over some six thousand persons employed by those representing the major stakeholders of that segment of roadways. By minimizing the power of those who govern, corruption will also be minimized.

The administration of the Federal Aviation Authority rules can be handled by an independent entity representing and paid for by the stakeholders in air traffic control. The Federal Food and Drug Administration could be replaced with a mutual insurance company where all producers and marketers of drugs underwrite all harm or damage to anyone as a result of ingesting their legally acquired drugs. This would include any side effects. A free market for human organs would end any queuing for transplants. Such a mindset can avoid the corruption caused by concentrated power. In these ways, great power can be diffused over thousands of non-governmental persons.

No government regulators may be controlled in any way by the controlee. Those who have been government employees may never be employed by controlled entities.

Government shall foster the creation of volunteer and charitable institutions to provide for matters of care and compassion.

The control of government must be limited to "left"-brain rational thinking. Government cannot successfully engage in "right"-brain matters of care and compassion, as these are without limit and beyond the capacity of government, and axiomatically require the involuntary participation of some to care for others—a form of slavery, as some sovereign persons become coerced into providing care for others. Because "right-brain" (caring and compassionate) thinking and "left-brain" (critical thinking) ability are generally uneven among and between the genders but equal among some, a "left-brain critical thinking quotient" would be helpful in the selection of persons for government. As we know now, or will learn eventually, care and compassion exceed the capacity of government to provide.

Corruption of the Legal System

All matters of government enforcement, matters of social and commercial intercourse, and matters of conflict among the citizenry depend upon a legal system. Experience suggests that such systems are all too frequently corrupted. A Free-orderly society must beware of the attorneys-at-law. The experience of ancient Rome, two thousand years later, and in the United States warns that lawyers can render a legal system ineffective. Attorneys-at-law, then and now, corrupt legal systems to serve their own interests, rendering the systems ineffective and, in some cases, worthless. Such behavior is normal human behavior. A successful system must be structured in a manner that prevents such normal human behavior from causing its debilitation. A successful Free-orderly society must style its government and its legal system in a manner that produces efficient, rapid, and effective adjudications. Special interest groups, including attorneys-at-law, must be prevented from corrupting the system to their own advantage. Lawyers who make the system of justice efficient contribute to the orderliness and productivity of the entire society. Attorneys-at-law who carve out a "piece of the pie" for themselves without increasing its size make little or no contribution to society. They frequently do this by making litigation cumbersome, inefficient, and protracted, and by co-opting damage awards.

Government Enforcement of Orderly Behavior

Government shall have the responsibility to mandate behavior among its citizens that is consistent with a Free-orderly society and to unwaveringly and universally enforce that mandate. This is to say that the supreme government shall establish and enforce laws protecting and preserving the sovereignty of each adult individual as they interface with each other, thereby producing the same standards of behavior throughout society. As most problems will be between sovereign individuals, a "People's Court" Judge shall hear all disputes between the sovereign parties; make a judgment; and remand discontented parties to a higher court, to arbitration, to special instruction and counseling, or to the State's Attorney. A Free-orderly society cannot tolerate individuals who intentionally harm the person, property, or minor children of others. It is not for the government to be concerned with why a person will not or cannot respect the person and/or property of others. These are matters of care and compassion and therefore are not and cannot be a function of government in a Free-orderly society. When such matters of personal behavior are unique to the States, Provinces, or Regions, their inferior governments should govern at that level within the standards here set forth.

Law Enforcement Agents are agents of the government. As agents of their employer, their employer is responsible for all of their actions when acting in the line of duty. If they fail to perform their duties in a manner prescribed by their employer, they are only subject to the terms of their employment contract.

Respect for the person, property, and minor children of others is primarily a result of inhibitions, inborn or inculcated in the very young. Some call these inhibitions a conscience. However engendered, those not possessing such inhibitions must acquire them if they are to remain in a Free-orderly society. For persons who do not possess such inhibitions by adulthood, their inhibitions must be acquired by employing their own rational processes. Such persons must reason that the consequences of deviant behavior are so undesirable that they will refrain from it.

Prisons are symptomatic of a morality that conflicts with a Free-orderly society. The concept that law-abiding citizens should have their production and property confiscated by government force and given to support criminals is incompatible with a free society. The question must be asked, "How is it morally different for the tax collector's agents to confiscate one's production or assets at gunpoint to support a criminal than for the criminal to have confiscated the same assets at gunpoint by himself for his own support?" This is not only in conflict with a Free-orderly society; it is in conflict with such a society's own moral standards. Societies with developed economies have begun allocating resources for the maintenance of individual prisoners in amounts that exceed the production of the average worker in the society. In fact, to incarcerate a young criminal for a life term without parole will cost the equivalent of the lifetime earnings of two average laborers. The providing of large amounts of resources for the feeding, clothing, housing, medical care, and entertainment of those who have proven themselves unworthy of a place in society has no place in a Free-orderly society. It cannot be accomplished without government coercion of citizens who behave in a manner that respects the person and property of all others. What kind of reasoning would grant a murderer or rapist a life annuity? *Life is precious, but not more precious than a Free-orderly society.* A Free-orderly society cannot enslave the law-abiding producers into the service of the criminal. Imprisoning an individual for life has nothing to do with behavior modification and is the product of a society that has placed a greater value on life than on freedom. Such has no place in a Free-orderly society. For the United States to bring thirteen Somali pirate murderers to the USA and award them the equivalent of a $1,500,000 life annuity (life in prison) disgraces the human intellect.

In a free society, jails serve to hold people for adjudication of an alleged crime. Those not possessing the inhibitions necessary to live in a Free-orderly society must either acquire them or leave the Free-orderly society. Historically, societies have imposed punishments to modify or correct the behavior of those who transgressed the rules and laws of society. Punishments served to change the behavior of many and to dissuade others from such errant behavior. Commonly, the most persuasive and effective punishments were harsh. What is known is that there are no two human beings who are "wired" the same. Therefore, the successful modification of one's behavior is different for every individual. This is a task for a panel of psychiatrists and psychologists, not judges, lawmakers, and other lawyers. This is extremely critical for first-time offenders. It is extremely important for the future benefits or costs to society. Most transgressions are acts of one person violating the sovereignty or well-being of another person's person, property, or minor children. These will most commonly be matters of Civil Damages settled in Civil Court. In such a proceeding, the person might be remanded to the jurisdiction of a Behavioral Modification Panel, which commands the powers of criminal penalties.

Second-time offenders demonstrate that the first effort to modify behavior failed. A second try would likely impose harsher punishments relying on humiliation, superficial pain, and threat of still harsher punishment through suspension of a harsh sentencing. Third-time offenders become a direct challenge to society. If a society chooses to be orderly and free, it must determine how many transgressions will be tolerated before a person is determined to be incorrigible. If a Death Sentence, suspended, fails to modify one's behavior, it is difficult to imagine what else might.

In a Free-orderly society, responsibility for behavior that respects the person, property, and minor children of others rests on each adult individual or the parents of the minor children. The method of behavior modification, therefore, may be selected by the offender and submitted to the court of jurisdiction for approval. Whatever method is proposed to the court must be to the satisfaction of the court, serve the dual objectives of modifying the behavior of the transgressor, and serve to be a deterrent to other would-be transgressors. There will be available methods for behavior modification that are yet to be discovered. Society must determine how many transgressions it will tolerate by any one individual before judging a person to be incorrigible. The consequence of exceeding that tolerance must be dismissal from society. The obvious alternatives are: either the incorrigible must emigrate to some other place in this world or be assisted to the next. There is no place for them in a Free-orderly society. Some are inclined to have penalties that avenge an injustice. Some have disguised "revenge" as "justice" or "held accountable." While revenge seems to be in the normal range of human instinctive behavior, it doesn't have rational stature in a Free-orderly society.

In a Free-orderly society, the costs of the transgressors' acts and of any behavior modification must be borne by the transgressors. There may be those who wish to voluntarily assist the transgressors to change their behavior or to emigrate. In any event, in a Free-orderly society, government cannot shift the burdens caused by the transgressor to citizens who conduct themselves in a manner that respects the person and property of others. The government may choose to inflict floggings on transgressors. Flogging is inexpensive, affordable by the transgressor, highly effective in behavior modification, producing appropriate inhibitions, and very beneficial in dissuading others from such errant behavior. Such inhibitions can permit many to remain in a Free-orderly society. Flogging can be administered in a manner that does not impair the future productivity of the transgressor. In any event, whatever punishments are imposed must achieve the goal of modifying the behavior of the transgressor, work to deter other potential transgressors, and the costs must be borne by the transgressor or others who freely undertake those costs. A term in prison could be an alternative only for those able to sustain the costs of such a method of behavior modification. This could occur where a fine is assessed against current legitimate assets or against future earnings, paid and applied to the costs of incarceration.

"Suspended" harsh punishments would be a great deterrent to repeat deviant behavior. Imagine how one's having a suspended sentence of a flogging and being required to periodically witness such events might modify one's future conduct. Imagine how a suspended death sentence might modify behavior. It is in the transgressor's interest that the punishment is effective, or they face being removed from the Free-orderly society. The only options for the incorrigible are, as we have said, the option to live in some other part of this world or to be assisted into the next. Many who oppose the death penalty can find fulfillment in moving the incorrigible to some other place on earth. Any other nation or society that objects to the killing of the incorrigible sociopath need only provide a visa, and the person can be delivered forthwith. It is likely, however, that there will be rulers of primitive nations who, for a fee or motivated by compassion, will admit these people to their lands. Barring that, there is but one option—execution. A return to the Free-orderly society by a condemned person who has left the jurisdiction will cause the person to be summarily identified and executed. Keep in mind that, for most, behavior that respects the person and property of others does not rely on a rational process but rather is a result of behavior inculcated in the individual during the earliest years of life.

Many have difficulty accepting a system of capital punishment because, in some instances, innocent people would be put to death. Universally overlooked is a system that kills tens of thousands of innocent women and children each year. It is the transportation system. Inexplicably, these deaths of thousands are tolerable to people who find the death of an occasional person at the hands of a very careful judicial system intolerable. The likelihood is that such opponents to the death penalty do, in fact, place life ahead of all other considerations, which, as we have shown, is incompatible with a Free-orderly society. As the Free-orderly society becomes ever more orderly, the incidence of such errors will all but disappear with the disappearance of crime.

An act that damages the person, property, or minor children of others demands restitution. Restitution with an effective penalty (penalties that effectively modify behavior) having been made, the matter is closed. Failure to make restitution requires the government to impose further behavior modification. No purpose is served by the extradition of those who, by their own flight, admit their inability to respect the person or property of others. Adjudication of their crimes can be accomplished in their absence; any possible restitution from any of their assets can be sought, and were they to subsequently return to the jurisdiction, a harsh penalty would be summarily imposed.

Once a society becomes orderly as a result of the uncompromising enforcement of the requirement that all participants respect the person, property, and minor children of all others, and the resulting strong family structures develop, the number and degree of transgressions will become negligible. The cultural shortcomings that encourage, or at least do not discourage, anti-social behavior will tend to disappear. As we have suggested, if there are very few crimes committed, the probability of being falsely accused would become negligible.

Punishment that does not modify behavior sufficiently for a transgressor to function in a manner that respects the person and property of all others is a totally worthless exercise. It is in the best interest of the transgressor to modify personal behavior so that he/she may remain in a Free-orderly society. Also, punishments serve to dissuade potential transgressors from similar behavior. Punishments too slight to accomplish this do a disservice to the potential transgressors, for they fail to modify behavior and thereby do a disservice to society as a whole. Punishments that are ineffective effectively condone crime. Here again, cruelty cannot be a concept applied only to punishment when evaluating the total cruelty in a society. A society must also consider the cruelty inflicted upon one citizen by another. Also, it must be kept in mind that in a Free-orderly society, the punishment must be affordable for the transgressor. The burden of behavior modification cannot be shifted to others in a Free-orderly society. Practically any transgressor can afford flogging or castration. Those who can afford other costlier methods of behavior modification may choose such methods, so long as they are proven effective.

Efforts to reduce crime by erecting impediments in the path of the criminal are mindless exercises. The installation of home security systems and better locks fails to stop burglaries. Gun registration and waiting periods fail to stop killers. Telling female joggers to carry pepper spray or hornet spray fails to stop rapists. Alarms on children fail to stop kidnappers. They are only an abrogation of responsibility on the part of the government to stop the crimes by dealing with the criminal. Impediments merely serve as additional hurdles for the criminal to leap. Only effective behavior modification of the criminal serves a Free-orderly society.

In the humanitarian socialist societies of the twenty-first century, concerns for the welfare of transgressors have spawned a tolerance for disrespect of persons, property, and minor children. A society that tolerates assaults on other persons and encroachments on the property of others, and assaults on minor children, will endure such chaos. A society that imposes ineffective punishments is a society that tolerates the disrespect of persons, property, and minor children.

It should be kept in mind that inhibited social behavior is best achieved during the development of a child. Inculcation during one's childhood removes acceptable social behavior from the domain of government. Failure to infuse or inculcate acceptable social behavior in a child is to fail as a parent. To rear a child who becomes a socially dysfunctional adult is, by its very nature, counterproductive. As anti-social behavior is bred out of the population, punishment intended to dissuade others from such behavior will become significantly less important.

The requirement that all persons respect the person, property, and minor children of others could lead the government to become involved in the minute delineation of personal relationships. For the government to interpose in such matters opens the door to the destruction of freedom. For this reason, it is essential that governments only become involved in acts against the person, property, and minor children of another, which are overt and intentional. Any and all such unintentional, accidental, casual, thoughtless, or careless matters or claims of damages among and between individuals must be resolved as civil matters either through mutual agreement, arbitration, or litigation.

Sexual assaults of a person not in a voluntary circumstance need only be punishable by castration and surgically imposed impotence, the costs for which the transgressor will bear. There will be, as a result, no second offenders, and, as a practical matter, when the punishment is widely understood by the general population, rape will be a very rare offense. Such will most certainly and effectively modify behavior. Few understand that the diminished libido removes the Need for sex, and it is no longer a motivation or frustration. As difficult as this may be for some to understand, those once addicted to nicotine and have become free of the addiction might make the comparison and understand that without the Need, it is neither a frustration nor a deprivation.

Sexual advances by and upon adult persons who are in voluntary social contact—commonly referred to as a date—are risks borne by the persons who so voluntarily expose themselves to such contact and cannot be construed as rape. If adjudicated as a forced sexual assault by acquaintances, in a voluntary circumstance, then punishment should be the same as for any other assault or battery. The reason that sexual assaults of persons in voluntary social situations are not to be considered rape is that the penalty imposed for rape is so harsh that it would impair social contact. The risk of a false accusation about the nature of the relationship and the irrational emotional motivations for such a false accusation would expose the innocent to grave injustice and greatly inhibit normal social intercourse among men and women.

Entertainment containing violence and disrespect for the person or property will likely be so abhorrent that it will have no market. It will be contrary to the very nature of citizens of a Free-orderly society, where respect for the person, property, and minor children is inculcated into each individual. Populations found the entertainment of the first-century gladiators and other gory entertainment so abhorrent that it all but vanished. The survival of bullfighting, boxing, gory television, and movies suggests that humanity has some fascination with that which is horrifying. If it were to be proven that there exists some human Need served by the macabre, then presentations of crime reenactments combined with the actual attendant gruesome punishments might serve both as meeting such beastly Needs along with reinforcement of such deterrents.

The Free-orderly society places the responsibility for each person's behavior upon one's. If one chooses to or becomes addicted to chemicals, the burden rests upon the addicted individual or any others who freely choose to concern themselves with such matters. In a Free-orderly society, such a burden cannot be transferred to others unwilling to accept it. Those who render themselves incompetent through drugs, by their own hand, should be allowed to die or recover at their own expense in a place removed from the orderly flow of society. Their wretchedness is of their own doing, and society need not be witness to it. Those who would find fulfillment in ministering to them, in a free society, will do so. No punishment can be too harsh, cruel, or inhumane for anyone who supplies addictive drugs to a minor without the consent of the parents.

Prisons will not be a part of a Free-orderly society. Those in prisons must be resentenced in a manner that will successfully modify their behavior. The hopelessly incorrigible will be sentenced to death. Many in prisons will not be considered criminals in a society where income taxes are no longer and where people are sovereign over their person, property, and minor children.

Government and the Family

Families are so important to the social structure that government intervention in the formation of the family is justified and required. Strong families, capable of being primarily responsible for all matters of care and compassion as well as the nurturing, rearing, and education of the young, enable a society to be free and orderly. Strong families that include grandparents and other relatives and friends make stronger individuals of all ages. Young adults having strong family backing become far more productive than they might have been without it. Middle-aged adults carrying the major responsibility for the family can commonly do much better with support—emotionally and financially—and with the additional wisdom that might be provided by their elders.

In a Free-orderly society, the family is the incubator for the subsequent generation. It is in this incubator that the inhibitions are developed, which enable the individual to function in a manner that respects the person and property of all others. Both the mother and the father make important, lasting, indelible imprints on the psyche of their children.

As it has been shown, children have no stature under the law in a Free-orderly society. The governance of the children in a Free-orderly society must be totally within the purview of their parents, as parents exercise their total responsibility and absolute authority over the children. As it has been explained, once a child is "permitted" to a family, any abridgement of the parents' authority and responsibility is an invasion of the sovereignty of the parents. Such an invasion of sovereignty is doomed to a never-ending encroachment and the destruction of individual freedom. Such an invasion of sovereignty also becomes an intolerable intrusion into the practice of their religion. Only where the parents are adjudicated incompetent may the state intrude upon this sovereignty.

The family is of such great importance to the human psyche and social structure that we have a chapter dedicated to it.

Government Intervention in the Family

In a Free-orderly society, government shall not intervene in family matters except to ensure that the family has the ability to provide care, nurturing, and education for each additional child admitted to a family. To procreate without the willingness and ability to care for and rear a child to adulthood must be stigmatized by the label **"Irresponsible procreation."** This is the most insidious form of child abuse.

Minor children can have neither rights nor responsibilities in society. Rights and responsibilities are inseparable and coextensive, so minor children cannot be held responsible for anything outside of the family. All responsibility for the minor children must rest upon the parents, and all rights and authority over the children must rest with the parents. Parents cannot shed their responsibility for their children and their children's behavior. They can do so no more than they can for the acts and behavior of their animals.

There shall be no Juvenile Courts. Unless the parents are adjudicated incompetent, the government shall not intervene beyond permitting the admission of the child to the family. To do so is to assert authority without assuming responsibility, or worse, asserting authority and shifting responsibility. To do so is to intrude upon the sovereignty of the parents. To do so is to intervene in the religious or ethical conduct of the family and thereby erode freedom of religion and introduce a pattern of encroachment by government into family life. To do so is to forcefully substitute the values and judgments of one person or a group over others. Any one of these events is incompatible with a Free-orderly society. The conduct of the family in a Free-orderly society cannot be within the purview of government.

The parents in a Free-orderly society must have and accept total responsibility and authority for their children's nurturing, care, education, and conduct. From conception, children live or die by the parents' will, judgment, and initiative, or lack of it. Children live or die as a result of the way a household is managed, whether or not any number of household hazards are managed. Children live or die as a result of the driving skills of their parents. They live or die in the manner they are placed in a bed. Children live or die as a result of the automobile privileges they are allowed. They live or die as a result of athletic activities they are encouraged to participate in. The intercession of the government into families' judgment, once accepted, is endless.

The laws of some religions do not allow the killing (aborting) of an unborn child. Some religions allow the killing of an unborn child in the first trimester or with conditions any time prior to birth. Some religions allow infanticide, while others grant the parents authority to kill a wayward teenage child. One must ask, "In a free society that espouses freedom of religion, whose religion is to be followed?" The answer must be "One's own." In a Free-orderly society, each family must be allowed to be governed by its own religious values and laws. In this respect, the family is autonomous, and it is always sovereign. There can be no government intervention that does not apply religious values and therefore impose one set of religious values upon others. However repugnant this may seem to some, freedom of religion can exist no other way!

Parents are accountable for the social conduct of their minor children. Penalties and punishments that result from the behavior of minor children must be borne by the parents. The concept of a "juvenile court" is totally incompatible with a Free-orderly society. By its very nature, it seizes the responsibility from the parents and usurps the authority of the parents. Children are not responsible to society. Only their parents (adults) can be responsible for society. Children have no authority in society. Only their parents (adults) can have authority in society. The courts can only hold the parents accountable. If a child behaves in a manner that fails to respect the person, property, or minor children of others, the parents are responsible. No juvenile court can let the parents shed that responsibility in a Free-orderly society. For the government to take responsibility for the child only shifts the responsibility to other, unwilling citizens. Think of having a "dog court" where the government protects the animals' "rights" and takes the authority and responsibility for the dogs' behavior away from the owners, and thereby shifts the responsibility to other unwilling citizens. In a Free-orderly society, parents must stand before the courts and accept responsibility for the behavior of their minor children.

The courts may choose not to punish or engage in behavior modification of the parents if the parents can demonstrate that they will impose a punishment or other plan that will successfully modify the behavior of their offending child. This behavior modification must be to a degree that will enable the transgressor to function in a manner that respects the person and property of all others. In a Free-orderly society, the rights and responsibilities of minor children can exist only within the family or their religious group.

Parents who demonstrate an inability to rear children, by abuse, abandonment, or intentionally taking the life of a healthy child, will be denied any future permit to rear children. Society can ensure this through sterilization. Parents who intentionally kill their healthy children are no threat to society. Such people constitute a rare anomaly. There are numerous cases where parents have killed their healthy children. Tens of thousands of children die as a result of the ineptitude of their parents. As has been mentioned, these ineptitudes include all manner of childcare and supervision.

These deaths have little or no negative consequences with respect to the survival or effectiveness of the society. However, governmental intervention in such matters has an enormous consequence on freedom and the sovereignty of the individual. Any encroachment upon such sovereignty of the individual becomes open-ended and a cancer in the whole body of individual freedom. The betterment of childcare in a Free-orderly society comes from basic human instincts, religious organization teachings, and the teachings of private organizations committed to that task. Parents who produce children who succeed in a Free-orderly society will populate the Earth, and those who do not will not. Humanity will be ever better as a result.

Where mindless, wanton, and egregious child abuse occurs, the question of sanity or, at the least, competency must be considered. Even if all parents are determined to be competent to rear children at the time they receive a permit for a child, all manner of illness or injury can render an individual incompetent. As with any other situation where a person is adjudicated "incompetent," a court of justice can have the authority for the determination of competency. In such a case, a guardian or guardians will be appointed, as with any other case of adjudicated incompetence. In such a manner, the guardians will be granted all authority over the parents, their property, and their minor children. The guardian or guardians at that point will have the sole authority to determine the fate of the children. Individual sovereignty for an incompetent individual is no more appropriate than is individual sovereignty for a minor child. This is a line at which society can intercede in the matters governing the family. Western societies have demonstrated beyond any question of a doubt that in the absence of such a strong line, the encroachment of government into the authority and responsibilities of the family is endless.

Abuse of children by parents presents a powerful dilemma for those who understand the nature and merit of the freedom of the individual. What constitutes abuse in the judgment of individuals varies greatly from person to person. For this reason, only the most egregious and wanton abuse can be considered when questioning sovereignty over minor children. All lesser abuses, mistreatment, or inadequate care of children that may be the result of the ineptitude of their parents are so arbitrary that there must be professionals to modify their ways of dealing with their children. In a Free-orderly society, where there is an understanding of the importance of rearing the next generation, the education of persons before undertaking child-rearing will become extensive and will minimize the mistreatment of children. This is discussed further on. If, however, the parents must be declared totally incompetent and a guardian given total authority over their person and property in order for the state to intercede, it raises the bar sufficiently for any judge so that a determination of child abuse could not be made over minor incidents. The reader must remember this is in a society where there will not be irresponsible procreation. It is a society of strong multi-generational families. And as a result, there will be little child abuse by parents, and in those few cases where there is, the children will be sought after by the many competent childless couples anxious to adopt them. If the parents, so adjudicated, were ever able to re-establish their competency, they would likely never receive another permit for a child.

If a father or mother was inclined to abuse their children, and the spouse was to stand for the defense of the children and be assaulted in the process, the aggressor must be subject to laws protecting adults from assault by others. The Family Contract between the parents would likely have provisions for such eventualities.

As the family in a Free-orderly society has total responsibility for the care and nurturing of its children, occasionally the family will be confronted with difficult decisions regarding a severely disabled or deformed child. Inasmuch as the responsibility rests with the family, the authority to make such a decision rests with the family. Such decisions must weigh the burden of maintaining a severely disabled child against the deprivation borne by the other children of the family. If religious leaders exercise authority in such a decision, it would follow that the religious group assumes a permanent and commensurate responsibility. In no event, in a free society, shall government intercede with either authority or responsibility. For the government to exercise authority in the matter is to encroach upon the freedom of the family and/or the religious group. For the government to exercise responsibility in the matter is to shift responsibility to citizens who would not willingly accept such responsibility. Either is coercion and violates the very essence of freedom.

If traditional societies are to evolve into Free-orderly societies, they must abandon mindless traditions and institutions of government that prevent the realization of a Free-orderly society and replace them with rational actions and institutions that serve the goals of a Free-orderly society.

In a Free-orderly society, when parents engage through contract other persons to provide supervision, care, or education for their children, the parents are primarily and totally responsible for whatever happens to the child or children while being provided for under the terms of the contract. Any harm or dissatisfaction with or from the services provided is a matter between the parties.

Government and the citizenry cannot be responsible for the consequences of such engagements, whatever the outcome.

Over some ten thousand generations, we have evolved in the incubator of the nuclear family of a mother, father, and siblings that make important, lasting, indelible imprints on the human psyche. Whenever this fails to exist, there exists a missing link in the person's psyche.

Adulthood

The concept of adulthood in a Free-orderly society is one of establishing the point in the development of a person at which society determines that a person is capable of assuming total responsibility for themselves (the ability to fend for him/herself). For purposes of law, the government must establish an appropriate age or method of establishing the point at which a child becomes an adult. Inasmuch as individuals mature at different ages and financial competence varies from person to person, the law might accommodate these differences up to a point. Once established, adulthood must confer total authority and total responsibility for one's. Adulthood creates another sovereign individual. This cannot be abridged in a Free-orderly society. This is the very essence of freedom.

In a Free-orderly society, prior to adulthood, a child is virtually the "property" of the family, and the family has total responsibility and total authority over the child. Adulthood completely changes this legal relationship of responsibility and authority. There can be no caveats or abridgements to adulthood, such as a greater age for the use of alcohol, etc. Such is arbitrary and a denial of the sovereignty that is conveyed with adulthood. Either one is totally responsible for and has authority over one's self and conduct, or one is not.

A child who becomes an adult but remains dependent upon the parents for financial support or maintenance can enter into an agreement with the parents. The agreement would establish what authority is to be vested in the parents, what are to be the responsibilities of the adult child, and what support is to be provided by the parents. It would be common for parents will be willing to support the higher education of an adult child and, in doing so, demand certain standards of conduct, performance, and responsibility toward the family. This would be strictly a contractual relationship between adults.

By introducing this flexibility into the matter of establishing the point or age at which a child is declared an adult, the age can be set at the earliest age at which a person can fend for himself or herself. This, by definition, means that the person granted the authorities implicit in adulthood and held by all other adults is determined to be able to accept all of these responsibilities. In different societies, the age may vary widely. At different points in the future, it will vary widely. In a primitive society, an age of fourteen might be sufficient. In a society where girls are married at thirteen, that might be the appropriate age for girls or just married girls. In a developed society where driving is a requirement for adults, sixteen could be an appropriate age. Where parents have little concern for their progeny, the younger ages may be appropriate. In a society where extended responsibility by the parents is desired, older ages will be sought. Free-orderly societies that demand that parents have extended responsibilities also grant the extension of commensurate authority. Parents may prefer a younger age, where they may shed their legal responsibility for any crimes committed by their children, but may enter into a contractual arrangement where they provide financial aid to them in return for a commitment to specific standards of conduct and behavior.

Immigration

A requirement for immigration into a Free-orderly society would be that immigrants must be able to assume financial responsibility for themselves and their immigrating families to meet the same standards as those for permitted procreation. The consequence of such a requirement would be to create a situation that was common in the United States in the nineteenth and twentieth centuries. One family member, commonly the husband or father, would immigrate, gain employment, and, when financially able, bring the remainder of his family to the United States.

With immigrant families having standards of living equal to at least the established minimums, immigration will be neither a threat nor a burden on a Free-orderly society or its citizens. This is to say that no one would be permitted immigrant status who cannot provide for themselves and/or their immigrating families a standard of living equal to or above society's minimum standard. This assures that the immigrants will be contributors to the overall well-being of the society. Such immigrants would likely have a long-term employment contract. Such contracts are likely to be restricted to those positions that cannot be filled domestically.

Immigration and the admission of temporary labor from foreign countries are not to be confused. The admission of temporary workers to accommodate a temporary demand for labor may, from time to time, be beneficial. Because of the vast disparities in wages from one society to another, a minimum wage may be imposed.

The supply of immigrant labor to fill temporary shortages has the effect of raising the standard of living of both the existing citizenry and the immigrant labor. A minimum wage is a subjective value. It must be an amount that will enable one adult to provide for himself/herself basic food, clothing, shelter, medical care, transportation, continuing education, taxes, and any other things that society deems essential. A minimum wage would be a justified impediment to the free market, as a free society establishes minimum requirements for the establishment of a family.

It would prevent an influx of labor from driving wages below any established minimum standard of living required for a "Family Permit." For the benefit of the economists, it could have some elements of elasticity. Apprenticeships may provide any one or more of these essentials in-kind. Apprenticeships may take the form of higher education provided under a contractual commitment to serve the provider for a time certain in a manner similar to that of the military academies.

In any event, in a Free-orderly society, immigrants are assets—contributors to the total production of the society. It is only in a Socialist society that immigrants are able to be liabilities.

Not-for-Profit Organizations

The not-for-profit organization brings together the elements of production, some of which are provided by donations. Charitable contributions of capital frequently do not seek or require market-rate returns. Charitable contributions of labor are given without the expectation of wages. Not-for-profit organizations, therefore, commonly intend to provide products and services at prices below what the market rate would be. Such organizations serve missions other than profits. For this reason, not-for-profit organizations should be chartered to function in areas that are not conducive to or competing with for-profit organizations.

Some not-for-profit organizations might have the mission of providing consumers with information about products and services. The risk of a profit-motivated organization providing consumer advice and information is that there will develop a conflict of interest or outright corruption will develop, as has been experienced. Such organizations commonly obtain their revenues from those they serve as fees or through subscriptions.

Enterprises that by their very nature have a "natural monopoly" can be better operated as a not-for-profit as long as that monopoly exists. A regional airport, for example, could be operated as a non-governmental not-for-profit. Landing rights, aviation services, and parking spaces can be sold at auction periodically. Capital contributed initially by the government can be valued on the basis of those revenues if purchased from the government. Subsequent acquisitions of land would be at market rates. Labor can be remunerated at market rates. From its initial non-governmental operation, the mission of such a not-for-profit is to balance the market forces, maximizing the economic benefit of the facility. Even if the government were paid the evaluation price for the facilities, its monopolistic position would require that it remain a not-for-profit organization. The same can be applied to segments of limited-access highways. Public schools can be sold to not-for-profit or for-profit organizations. The proceeds could be used to subsidize student tuitions, perhaps through a voucher system, as the educational system evolves from government responsibility.

When assets now owned by the state are not distributed through the National Fixed Investment Trust, they could be sold or donated to non-profit entities to be managed by directors or trustees representing the interests that the asset serves. For example, a section of federal highway could be sold or donated to a not-for-profit corporation, the directors of which would be appointed by the contiguous county and city governments that it serves. Also represented could be the major commercial interests that it serves. The diversity, limitation on terms of office, and number of interests represented would minimize natural tendencies toward corruption or the serving of someone's special interest. The diverse special interests would tend to optimize the capacity of the highway and its cost to the users through tolls. The funds required for the purchase of a facility from the government can be financed by bonds supported by the revenues that the facility produces, or by a loan from the government entity from which ownership is acquired.

Educational institutions have long traditions of serving missions other than profit. The risk with these and all other not-for-profit organizations is that the trustees, directors, commissioners, managers, or faculty will corrupt the mission by diverting funds to their own benefit, which exceeds that to which they would otherwise be entitled in the free marketplace. This will take the form of excessive salaries, benefits of all types, including pensions, and contract services provided on a non-competitive basis by "insiders." Only the organizational structure through a restrictive charter can minimize these abuses. Representation of the interests of all stakeholders, limitations upon terms of office, and significant numbers of those who serve on each governing body will succeed in minimizing corruption in a competitive arena.

Experience shows us that those who wield governmental authority are inevitably corrupted directly or indirectly by those whom they should govern. Every board, commission, or agency will ultimately be infiltrated by those whom they are supposed to govern. The industry or group that the bureaucrats are to govern has the greatest interest in controlling the composition of the board, commission, or agency. It is for this reason that the solutions to the governing of any economic activity must be relegated to the laws enforced by the courts or by mutual insurance and bonding companies.

Government agencies such as the Federal Drug Administration could be converted into mutual insurance companies. Run by a board of diverse interests, the revenues of the insurance company would come from premiums paid by the drug companies. The monetary liability for damages would be detailed, specific, limited but generous, indexed, and specified in the charter granted by the government. Failure to satisfy these requirements would expose the insurer to unlimited damages. Without a durable waiver of total responsibility, physicians would prescribe only insured drugs, but sovereign individuals could have access to any drug by executing a durable waiver and accepting the entire responsibility for the outcome.

Such an insurance company, that insures the product, should undertake such matters as the inspection of meat.

Consumers are responsible for their own consumption. If they want inspected meat, they should buy only insured meat – inspected and insured by a company holding a respected rating. Both the insurance company and the rating service are private entities. In many jurisdictions, such a system of responsibility would be far more reliable than one depending upon a corruptible bureaucrat.

Also, other traditional government regulatory functions such as investment and securities industries, and a whole range of professional regulatory bodies, can be replaced by not-for-profit insurers. The Federal Deposit Insurance Corporation functions largely in this manner.

There seems to be a general lack of effort to conceive non-governmental solutions to problems. Many such non-governmental solutions and ways of accomplishing things can be found if only they are sought.

Government schools can be sold to not-for-profit companies that would provide for publicly funded students through vouchers. Such could be funded by bond issues that would provide the school boards with money to provide vouchers. Public schools could be donated to the not-for-profit entity and operated in the same manner.

Such "private" schools would compete with all other schools, providing choice to the parents and the environment for true innovation in education. One needs only to look back at the monopolistic phone companies and their product and ask, "Would the phone system look anything like it does today if those monopolies were still in force?"

Government Ownership

In a free society, the government may hold, own, and maintain certain monuments, parklands, and preserve lands determined never to have economic utility. If, ancillary to the function as parks, monuments, or preserves, there is an economic function, the rights to such use must be leased at advertised public auction at the minimum intervals that will permit amortization of the required capital investment. Preserved lands might result from a determination that a forest region is necessary for the well-being of the environment. A preserved land might exist as a result of a determination to provide habitat for some animal species. If property rights such as air transmission frequencies are created by the government and are not granted permanently, they should be leased at auction at intervals that will permit the amortization of the capital assets involved. When the state retains title to land, the attendant requirements and caveats enumerated above must be respected.

It might be justified for the state to undertake large public works projects with large risks, which private capital would be unable to assume. The government might undertake large dam projects, unusually difficult highway construction projects, and the like. If, after completion, such projects or any portion thereof had a positive cash flow, it or that portion could be put into a for-profit corporation and distributed to "National Fixed Investment Funds," enabling the earnings of the asset to flow to the consumers. This can be done with any government ownership in profitable enterprises. Otherwise, such activities are fraught with the danger of graft, corruption, and waste. It may be determined that the projects might be justified for defense, but the risks are the same.

There is no deed too extreme in the defense of a Free-orderly society.

Among advanced countries, intellectual property rights are the targets for violation. A society that wishes to protect its intellectual property rights will need to expend great effort because its rapid technological advancement will make its intellectual achievements the envy of all others. Not as forthright as armed conflict, the defense of intellectual properties will require adroit diplomacy.

Defense of a Free-orderly society includes threats to society that will be meteorological, seismological, biological, and astronomical. A government may be involved in defense against and recovery from severe weather; the effects of earthquakes; destructive asteroids, solar storms, pandemics, or other celestial threats of which we are yet to learn. Governmental assistance in matters of survival of the citizenry resulting from natural disasters can include the economic restoration of the damages. In a Free-orderly society, most losses will be addressed by insurance, charitable institutions, and gifts.

Defense of society against infectious diseases is a natural function of government. Defending a society against threatened or real assault from whatever the source must be a threat to the whole society. National government involvement in any regional problems that do not affect society as a whole result in one region, area, or group being protected at the expense of another. Such becomes the source of all manner of political mischief. If states have the national government pay to maintain communities on barrier islands at the expense of populations in the center of the continent, such losses must be ameliorated through insurance or other group combinations formed to produce reserves to serve specific losses.

Foreign Relations

Foreign relations might include extradition for the restitution of persons who have done harm to the person or property of citizens. Foreign relations include international property rights and international commerce. Foreign relations are critical to the defense of a nation. And foreign relations may even include the purchase and holding of preserved lands in other countries. One example might be jungle lands held and preserved for the benefit of the earth's atmosphere by a wealthy nation within the borders and subject to the sovereignty of a poorer country. Foreign relations will include environmental matters that cross borders or matters governing Earth's indivisible assets. There are many foreign matters that are of critical importance to the security of a Free-orderly society. Joint scientific projects advance the highest essence of mankind. Meddling in the affairs of other nations can only be justified if doing so is absolutely essential to the survival of the Free-orderly society.

The entire human experience has not produced a government that has created and preserved a Free-orderly society. The Athenian democracy that produced a period of freedom and unparalleled achievements succumbed to corruption by the electorate. Plato and Aristotle lived in a failed democracy. Each searched for an alternative. Aristotle's solution was an inclination toward a republican form of government. Concepts that are embodied in a republic influenced Rome's republic and intellectual discourse for centuries. But the Roman Republic became dysfunctional in the first century BCE as a result of corruption and strife among the social classes. Some seventeen hundred years later, contributions by John Locke were timely enough to be combined with those of Aristotle and others to influence the drafting of the United States Constitution. Intended to be a republic of limited government, the founders' intent was dismantled by the Amendments, so the United States succumbed to the muddle and tyranny of an intrusive, corrupt democracy.

The Western democracies, like Athens two-and-one-half millennia earlier, are hopelessly and fatally corrupted. Virtually all voters sell their votes — some for as little as a pack of cigarettes. Most sell for much more. For whatever amount, it is hardly different than the patronage of the Roman senators and the corruption it bred. Consciously or subconsciously, each voter seeks some personal benefit. Some desire greater direct payments from the government. Some seek indirect benefits from government programs that will improve their personal economic condition. Some seek government-erected impediments in the marketplace from which they will benefit. These might take the form of enabling the voter to sell his labor for more money and less effort at the expense of his fellow consumers. Or it might take the form of restricting competition for the businessperson in order to widen profit margins. It might take the form of compulsory widening of an insured risk group to force the consumers having the least risk to pay for the burden of those having greater risk. Some voters are motivated by instincts of Care and Compassion that would have the government force some of the population to care for others of the population. This fulfills nearly all the elements of Slavery. Obviously, voters paying the taxes for these benefits vote for those politicians who promise to reduce their tax burden.

The corruption that most voters recognize is the direct and indirect payment by special interest groups to influence politicians and elections; to cause the politicians to erect impediments in the markets that benefit them; and/or to exempt them from legal burdens that impair their competitive position. Commonly, these are tax burdens. It is this type of corruption that exposes the magnitude of power that the politicians have granted themselves, and which is most visible and to which the voter is willing to admit.

The Western democracies take comfort in the statement that "democracy is the worst form of government except for all of the others." Those willing to look beyond this statement will see that even the Western democracies are fatally flawed and cannot endure. It is for this reason that alternative forms of government must be sought. Virtually all voters have a prejudicial interest in the system to the point of dysfunction. None is likely to relinquish his/her special interest until the democracy serves virtually no one.

At about the time our original 13 states adopted their new constitution, in the year 1787, Alexander Tyler (a Scottish history professor at the University of Edinburgh) had this to say about "The Fall of the Athenian Republic" some 2000 years prior:

"A democracy is always temporary in nature; it simply cannot exist as a permanent form of government. A democracy will continue to exist until the time that voters discover that they can vote themselves generous gifts from the public treasury. From that moment on, the majority always votes for the candidates who promise the most benefits from the public treasury, with the result that every democracy will finally collapse due to lose fiscal policy, which is always followed by a dictatorship."

The average age of the world's greatest civilizations from the beginning of history has been about 200 years. During those 200 years, these nations always progressed through the following sequence:

- From Bondage to spiritual faith
- From spiritual faith to great courage
- From courage to liberty

- From liberty to abundance
- From abundance to complacency
- From complacency to apathy
- From apathy to dependence
- From dependence back to bondage

There has been little or no experience since 1787 that denies those observations. Looking at where our society would rank on this list, the prospects are rather ominous. If this sequence were to be repeated, one could hope that, subsequent to the period of bondage that follows the collapse of democratic socialism, there would be a brief period followed by "spiritual faith" in a Free-orderly society, which is described here.

The Cost of Government

Wherever possible, the costs of government must be borne by those who use its services. For example, those who use the court system should bear the cost that they impose on the system. If the government were to be involved with matters relating to motor vehicles, then all the costs of such involvement should be borne by those who own and operate motor vehicles. The cost for the government to administer matters relating to the formation of families must be borne by those who form families.

The functions prescribed here for the national government are basic. All citizens benefit from them. All citizens must bear the costs of providing these functions. Where the government provides for the enforcement of that human behavior that respects the person, property, and minor children of all others, all benefit. Where the government provides for private property rights and for a free, open, honest market economy, all benefit. Where the government provides for the defense and preservation of a Free-orderly economy, all benefit. Where the government provides for Preserve Lands, all benefit. As we have mentioned, taxes placed upon consumption are borne by all and in an amount related to the amount of consumption, which bears some relationship to the benefits received by each consumer.

To emphasize the point again, any such consumption tax must be levied on all consumption equally. Any discretion in how or on what services or products will be taxed will produce the rankest corruption. The benefits of any exemptions will produce such great monetary benefits to some as to be capable of corrupting almost any government official. We have more to say about the disparity of wealth and income further on.

Government and Dissent

In a Free-orderly society, government shall tolerate any and all intellectual dissent expressed within a building, structure, or enclosure of any kind, with the exception that the expression or promulgation of ideas that incite people to act in a manner that fails to respect the person, property, and minor children of others cannot be tolerated.

Those who would conduct demonstrations in the streets generally do so with the intent of applying coercion. Coercion is absolutely an intolerable activity in a Free-orderly society. Street demonstrations would be tolerated in a Free-orderly society only if they can be shown not to coerce anyone. Compared to early times, present-day means of communication are so varied, ubiquitous, and ever-expanding that exchange of ideas is assured without street demonstrations that coerce and disrupt a Free-orderly society. Here, we should address free speech. Free speech is fundamental to a Free-orderly society. The free exchange of ideas is the road to maximizing human progress. Human experience has demonstrated that constraints on the expression of ideas, however repugnant to some, restrict human progress. The problem is that such evaluations of speech are commonly arbitrary and prejudiced. But speech must not cause real or tangible harm to others who are obeying the codified and moral laws of society. Because a Free-orderly society is preeminent, speech and writings that incite people to destroy freedom or the orderliness of a free society cannot be tolerated. The very act makes them an enemy of a Free-orderly society. Such persons must reside in some other society, if in any at all. If a Free-orderly society is not the paramount end, it will be vanquished by a superior nation.

Government and Religion

In a Free-orderly society, government shall not be involved in matters of religion or religious belief so long as they are practiced with respect for the person, property, minor children, and religious practices of all others that can be practiced in a like manner. Each religion practiced in a free society has this right and responsibility. To preserve that right, each must respect the rights of all other religions. No legislation would be permitted that would impose a religious belief or value on others, nor would it be permitted to require others to voluntarily submit to those values. Religious leaders must concern themselves with teaching and imposing their beliefs on their own followers, doing so without the assistance or the coercive power of government.

Government in Matters of Care and Compassion

Most social maladies that governments have attempted to cure, such as hunger, poverty, and illiteracy, have a root cause. That cause most frequent, in the industrialized countries is **Irresponsible Procreation**. As so often is the case, only treatment of the most visible symptoms of the social malady is attempted. So long as irresponsible procreation is tolerated, these societal ills will continue to be a plague.

In a Free-orderly society, government shall not provide or be involved in matters of the "care and compassion" of its citizens after a specific period of transition to the Free-orderly society. The government of a Free-orderly society cannot provide care and compassion **for** some citizens without encroaching upon the sovereignty of other citizens. It can only do so by exercising its powers of coercion—the very act as such is anti-freedom. For the government to use its coercive taxing power to provide care **for** some of its citizens is to use its coercive power to enslave a portion of its population—at least part-time—in the galley rowing the government's ark of compassion. In order to provide care, the government must confiscate the production and assets through taxes from one class of people to provide care for others. To say it another way, the government must enslave one class to serve another class. It enslaves those able to work and produce to service those unwilling or unable to work and produce. This was expressed in a slightly different way by Karl Marx: "From each according to their ability and to each according to their need." This experiment was tried in Eastern Europe and failed in a span of seventy years. It has been tried in Western Europe and the United States for seventy years in a slightly different form, and it failed. For a society to enslave the productive to serve the unproductive is to navigate a course having neither direction nor destination. Further, such activity by the government totally fails to respect the property rights of its citizens. It further fails to respect the person of its citizens when it threatens bodily harm if they fail to comply with the taxes so levied.

The demands for care and compassion are so boundless that even total enslavement of the productive population could not satisfy them. It follows that care and compassion must be rationed, and in a free society, this rationing falls only to those who freely and willingly accept the responsibility of providing it. Governments that try to provide care and compassion will ultimately fail, as human needs for care and compassion, real or perceived, are without limit and vastly beyond any government's ability to successfully deal with.

One might be inclined to ask: "Is it more logical for the production and assets of the population of Alaska to be confiscated at the gunpoint of the revenuers and given to the people of Puerto Rico than it would be for the production and assets of the population of Sweden to be confiscated at gunpoint and given to the people of Ethiopia?" Can that be reconciled with respect for the person and property of others? Patently, it cannot.

Those of a Free-orderly society who are concerned about matters of care and compassion should spend their energies and assets providing it. They should not spend their energies and assets in an effort to persuade and corrupt the government to use its coercive power to violate the person and property rights of others to serve their own mission of providing care and compassion.

It must be kept in mind that when the government takes responsibility for the care of its individual citizens, it is—or will eventually be—granted commensurate authority. Inasmuch as the needs for care and compassion will vastly exceed the resources of any economy, such commensurate authority is incompatible with a free society. When the government assumes responsibility for one's care, it implicitly or explicitly assumes the authority to administer it. This means that if the government is responsible for one's healthcare, it has the authority to require automobile passengers to wear seat belts, to require cycle riders to wear helmets and other protective attire. Using the same reasoning, it can even require that one wear a helmet when traveling in any vehicle. Further, by the same logic, the government can dictate that one does not smoke and what one will and will not eat. To be obese is to eventually be a burden upon other citizens in a society. It is manifest that the concept of government being assigned or assuming responsibility for the care of its citizens is incompatible with a free society.

In a Free-orderly society, care and compassion must be the responsibility of the family. It should be of great concern to the religious groups and further served by volunteer organizations that freely choose to concern themselves with such matters. But because the needs for care and compassion are unlimited, provisions for care and compassion must be rationed. Only those willing to freely accept the responsibility of providing care—the family, the religious organizations, or other charities—should make the difficult and sometimes harsh decisions about its allocation. Unproductive individuals must receive their care and support from their own reserve resources, their family, or others who freely and willingly provide their support. Institutions of insurance, along with the church and other charitable institutions, must be relied on to assist families in providing whatever care and compassion family members will receive. The hopelessly infirm survive either from their accumulated resources or from the voluntary care of those motivated by compassion.

Life should not only have a rational beginning but, where and when circumstances permit, it should have a rational end. The government should erect no impediments to restrict a competent individual from documenting the terms and conditions for the termination of one's own life. Whatever is the will of an individual regarding his or her demise is not a matter of concern for the government. Too frequently, governmental judgment about such matters becomes the enforcement of a particular religious doctrine. In a Free-orderly society, each individual must have the authority to follow his or her religious beliefs or his or her personal judgment. As long as the matter affects only the individual or his/her minor children, no government involvement can be exercised in a Free-orderly society.

Accepting these concepts of government involvement in matters of care and compassion, one can accept a role for government as society makes a transition to a Free-orderly society. The weaning of the population off dependence on government for economic survival may take some period of time. Prior to a time when this dependence is diminished to a point where eleemosynary institutions are capable of providing a safety net for those who do not have the ability to provide for themselves, a government-provided sustenance or supplementary income could be justified. The progeny of responsible procreation and competent immigrants capable of providing sufficiently for themselves and their families will cause an ever-declining dependence to a very low point where the natural instincts of charity will be capable of **providing** for these needs.

Healthcare

Dealing with the matter of healthcare in a Free-orderly society, all citizens must recognize that healthcare is not free like the air we breathe, that it cannot be equally available to everyone, and therefore must be rationed. This rationing will be by queue, quality, and/or price. As with Education Subsidy and Personal Saving, society may choose to subsidize Healthcare via Value Added Tax. To this end, the government may provide a partial, personally administered subsidy that is designed to assure Market Discipline for healthcare. A subsidy for healthcare is justified—at least in the mind of the author—by the fact that human longevity is critical to human achievement. This writing is a product of inordinate human longevity resulting from a most expensive type of healthcare, which, without subsidy, would not have existed at that interval.

A description of such a healthcare system is appended below.

Safety Net

When providing for these needs during a transition to a Free-orderly society with a "safety net," the benefits should be vouchers limited to subsistence ration supplements, administered in a manner that will not create disincentives for the recipients to become ever more productive. For the government to provide specific benefits in-kind or in money, for food, housing, or whatever, has only encouraged corruption and fraud. So, vouchers designated for the specific essential categories shall be spendable by the recipient to provide the benefits.

For the time subsequent to a transition period, a social safety net funded by the government would also be paid for by the consumption tax. Inasmuch as it provides all people assurance from the capricious nature of misfortune, depriving them of sufficient sustenance for survival, the burden rests on everyone. Basic sustenance includes clothing, food, shelter, transportation, education, basic healthcare, and communication. Each subsidy should be a specific ratio quantified by the prices in the marketplace of residence. The new institutions offered here will eventually provide a living income for all.

A Solution for Drug Addiction

To the extent that drug addiction is deemed detrimental to a Free-orderly society—its functioning and achieving its goals—its participants may be removed from society. Because the drug robs a person of his/her freedom, in a society where Free-orderliness is valued more than life, persons purveying illegal drugs shall be sentenced to death, "Suspended." A second conviction would only cause the suspension to be lifted.

The mentality of government functioning as a guardian for its population has fostered a most destructive societal disaster. The designating of specific addictive drugs as illegal has created a rancid "underclass"; it has filled prisons with people who have committed victimless "crimes"; over a period of four decades, it has frittered away billions of dollars without any success in achieving its intended goals, and it is directly responsible for the death of thousands in several countries. An understanding of personal responsibility and individual sovereignty over one's body shows that individuals are free to do with their person whatever they choose, so long as it does no harm to others. But those whose addiction adversely impacts the lives of others may be isolated—perhaps in an "Addicts' Colony." Once convicted by a court of law for a crime induced by the addiction, a person shall be sentenced to residence in an Addict Colony for a limited specific time—perhaps six months—and fined the cost of residence. Intentional possession of a banned substance should be a crime. Addict Colonies would provide basic food, clothing, shelter, education, healthcare, and transportation funded by the fines. At the Colony, the residents will be able to choose to live either in a drug-available community or a drug-free community.

Addictive drugs will be available at cost through dispensaries operated by a government or eleemosynary agency that operates Addict Colonies and provides drugs to maintain the addiction. The Agency will purchase the addictive drugs from the lowest-cost supplier nearest the point of production, meeting established standards. This will avoid the destructive consequences of illegal drug trafficking. By making drugs legally available to addicts and making the "cost" of illegally purveying drugs so high, illegal drug trafficking will all but disappear, ending the multiplicity of bad consequences of the illegal drug trade.

Healthcare would include counseling. Once addiction-free for some period established by a committee of therapists for the individual, he/she would be released from the Addict Colony. To the extent that the cost of residence in the Colony exceeds the fine, a lien against any assets of the resident, calculated at cost, will be generated. Upon being released from the Colony free of addiction, all court records of the incident shall be expunged. Society is better served by persons free of such impediments.

Families, schools, religious groups, and others should preach and teach the perils of even experimenting with such substances, for to fail in this effort is to fail in all others regarding the rearing of a meritorious youth. Humanity's best hope lies in its success in finding that area of the brain where addiction resides and developing a way of modifying it. *(Note: the sentence was duplicated; one instance has been removed.)*

The Socially Dysfunctional

Those who become "Socially Dysfunctional" may be required by the courts to be residents of "Communities for the Socially Dysfunctional." Such individuals will necessarily be harmless; they will likely be homeless; they will likely be unemployable; they will likely be addicted to some debilitating substance, and will be receiving Safety Net care.

The Community for the Socially Dysfunctional will likely be an abandoned military base, an abandoned prison from which the bars have been removed, or perhaps a farm-like community. These communities will be operated by charitable or religious organizations—by those who are compassionately inclined. Those who choose may become rehabilitated and return to a functional society.

Equality, Fairness, and Justice

The concept of "equality" in a society is poorly understood. The initial statement in the American Declaration of Independence that "We hold these truths to be self-evident that all men are created equal," has become widely misconstrued. It referred to "that they are endowed (equally) by their Creator with certain unalienable Rights—these are Life, Liberty, and the pursuit of Happiness." This statement was a declaration to King George denying the "divine right of kings." Each adult individual is a self-contained, mobile organism. To that extent, we are all equal. What is self-evident is that all are created unequal with respect to their physical and mental abilities, also with respect to their individual initiative, energy, and personal discipline to forgo immediate gratification for future pleasures. Every individual has a different intellect, the variety of which is endless. Every individual has a different motivational structure with which they develop his/her unique abilities. Each person has a different initiative or work ethic. Any social order that attempts to make equals out of those who are unequal must engage in the rankest discrimination and coercion. Compounding the inequalities of personal and individual differences are the inequities of good fortune and misfortune. For this reason, all who live in a Free-orderly society must accept great inequities in wealth, position, and accomplishment among the population.

Equality under the law is an intrinsic characteristic that cannot be compromised in a Free-orderly society. No one can be above the law that receives its validity from a society of sovereign individuals. Therefore, no law can be valid that provides advantages for some or disadvantages for others through the impairment of their sovereignty. This should be particularly observed when those making laws are inclined to favor themselves with either benefits or exemptions from obligations under the law.

An understanding of the social order presented here provides confidence that nearly all families will have an income above that which has been determined to be a minimum level. Or to say it another way, the minimum standard for the formation of each family establishes a standard of living that is acceptable to society. For those who fall below that level as a result of circumstances beyond their control, many private charities will stand ready to assist them. Where there is no irresponsible procreation, the incidence of families falling below the minimum standard will be low, and private charities will be capable of dealing with their requirements. The human need to be caring and compassionate is so strong that vast resources will be made available from private initiatives to families with sub-marginal incomes. No social order will ever provide for all the needs (and wants) of all its members.

The capriciousness by which life's good and bad economic fortune falls upon human beings is outside of any rules or laws. Therefore, the concepts of "fair and just" cannot and do not apply to good or bad fortunes or, if you prefer, good or bad luck. Society cannot control the manner in which good or bad fortunes fall upon the population and, therefore, cannot be responsible for providing "fair and just outcomes." There is nothing in the laws of nature that demonstrates what is fair and just. They are human concepts that, while applicable to interpersonal relationships, are not within the capricious nature of events that affect the lives of individuals. The government cannot convert such capriciousness into fair and just results. Any attempt to do so would be inconsistent with a Free-orderly society. Many private institutions are dedicated to ameliorating the negative consequences of misfortune on individuals. Many more will be created when governments disengage from such activities.

"Fair and just under the law" is related to equality under the law, as law must provide for any concept of freedom. A law that provides for freedom must, by its very nature, be superior to all who administer it. Therefore, no person under the law can be immune from it. To be so is to be freer than others. Here is where equality under the law becomes judged "fair and just" or "unfair and unjust." It is not uncommon for those who make the laws to make them "fairer and more just" to themselves.

CHAPTER 6

THE FAMILY IN A PROSPEROUS FREE-ORDERLY SOCIETY

If a society has any value, to preserve that value, it must be continued. That places procreation and the development of its children at the top of any priority and significance. It is in this endeavor that most individuals can find their self-worth. If humanity has any value, there are few occupations or professions that are superior in importance to rearing the next generation.

The family is the most fundamental social, economic, and educational institution. The family is the incubator, crucible, and mold of each human individual. Our brains have evolved over ten thousand generations in this structure. We have learned that both the organic development of the brain as well as the programming of the brain is irrevocably shaped by and during the nurturing years in the family structure.

It has never been successfully replaced, nor is there any prospect of it being replaced. The family is the fundamental building block of a *Free-orderly society* upon which all other institutions of a *Free-orderly society* rest. In the following paragraphs, we explore the nature of adaptive changes that will characterize the family organization in a *Free-orderly society.* Many do not desire to rear children. Such people may find their fulfillment by contributing to the rearing of children outside of a family structure. That is to say, people who do not rear children may directly be involved with their rearing through such activities as working for the educational or medical institutions that directly serve the rearing of children. Or indirectly, all manner of commercial activities can have a beneficial impact on the rearing of children. The wealthier a society, the greater will be the optionality for the better development of the children.

The family need not be restricted to blood relationships. Nor must it be restricted to one household. Its interests may extend to their common social, economic, and spiritual interests, as well as their security. Most commonly and fundamentally, the focus will and should be on raising and educating the next generation. It is likely to be a mutually empathetic, voluntary relationship that transcends gender and generations.

Until such time as it has been proven to the contrary, a *Free-orderly society* must accept that the family is the fundamental building block of any society. It is where each individual develops the self-image that governs most aspects of adult life. It is within the family that each successfully functioning adult has been, as a child, inculcated with the behavior of conducting oneself in a manner that does not harm the person and property of any other person.

As dictated by the evolution of the human species and traditional religions, the *"core family"* consists of a father, a mother, and children. Experience shows us that each parent plays a vitally important role in human development. No one has successfully denied that. The experience of societies that have ignored this fact has produced stark evidence of the validity of this statement.

Without relying on the mores and folkways of the past, it is becoming ever more apparent that parents have an indelible impact on the child. From some time before birth and during the very earliest months and years of life, the child develops his or her own identity from his or her parents (particularly from the parent of the same sex) and from their relationship to each other. Many in Western societies choose to deny the importance of the two parents in the development of a child into an effective adult. Such rationales generally are to accommodate personal needs, situations, or political expediencies. Single motherhood, divorce, homosexual lifestyle, or whatever irregularity notwithstanding, no society has as yet found a better developmental environment than that of two competent, natural, loving, and caring parents, or when strengthened even further by other committed and supporting adults of both genders.

In a free society, all other persons who choose and are accepted to be a part of a *"greater family"* may be freely appended to such a family and be dedicated to the mission of that family—the economic welfare, social development, education, and general nurturing of its children. This could include grandparents, great-grandparents, aunts, uncles, and friends committed to the mission of nurturing the children. It is manifest that the strongest families will be those who include the most relatives and friends of all ages serving the mission of rearing the children. As longevity increases beyond the childbearing years, multi-generational involvement with the rearing of children will be more important.

To procreate without the willingness and ability to nurture and rear a child is to assume an authority without the commensurate responsibility. This **Irresponsible Procreation** is totally incompatible with a *Free-orderly society*. Each partner of the **Family Contract** will have attained some minimum level of competence in order to qualify for a **Permitted** *child*. This will require certain educational achievements for the skills of a functional family, the bearing and nurturing of children, and the occupational skills necessary to support the family. If reverence for human life is to be preserved, conception and birth must be looked upon as the reward for years of self-denial and forbearance in preparation for the joy of competently nurturing, rearing, and educating a child for a productive role in society.

While government in a *free, orderly society* should not be involved in the conduct of the family beyond its formation, government must have an uncompromising role to assure the sound foundation for the life of each child. Traditional religions, for thousands of years, successfully established families that were sound building blocks and foundations for social order. For whatever reason, by the end of the twentieth century, the family throughout much of the world had crumbled.

Beyond the assurance that the commencement of a family and each additional child to a family has a sound beginning, any government intervention is a gross intrusion upon the freedom of the individuals in the society and, inevitably, a gross intrusion into the freedom of religious practice. The challenge to religious groups, after having received assistance from the government to deliver families that have a sound inauguration, is to build upon this and to help develop sound families capable of functioning successfully in a *Free-orderly society*. The family can then practice the values of its religious teaching in achieving the religious group's mission. The nature of this mission will be discussed further on.

The Family Contract

Marriage has evolved into a contract far different from what it was in earlier times. For millennia, there was a presumption that marriage signified the commencement of a core family with the commencement of sexual activity. It has evolved to a point that marriage may have no relationship to rearing children or sexual relationships. It has become a legal relationship, a contractual relationship imposed by law between two people serving the self-interests of each—a contract by law to provide financial benefits and/or emotional security. As long as such relationships do no harm to third parties, they should be of no concern to governments. They are appropriately arranged by contracts between the parties.

The marriage contract has become inadequate to fulfill society's need to provide for the bearing, nurturing, rearing, and educating of children through sound family structures. It has been ignored by couples who have established households with children without the benefit of a marriage contract. Successful *Free-orderly societies* will be those which require a *Family Contract* into which a sovereign man and a sovereign woman—and perhaps others—enter, agreeing to subordinate their personal interests to that of the nurture, education, and care of their children until adulthood. The goal, of course, is to produce good, productive citizens. The *Family Contract* presumes that the parties intend to rear children. In a *Free-orderly society*, all other persons may enter into any other personal relationship contract that they wish. This will likely replace government marriages and the marriage law. The process would be one that, at the termination of marriage laws, all existing marriage contracts would be considered to have, through their verbal consent, incorporated all the then-existing marriage law, which would be so enforceable. All subsequent *Family Contracts* would be governed by their expressed terms and conditions. Marriage would be essentially a religious event embodying whatever the participants choose to agree to. The most successful societies will hold that, without such a commitment to self-sacrifice, expressed in the form of a *Family Contract*, no one will be *permitted* to commence a family or to bring additional children into a family. In order to prevent *irresponsible procreation*, the government of a *Free-orderly society* shall require a *Permit* to be issued to a man and woman who have entered into a *Family Contract* and who demonstrate sufficient competence and will to physically, mentally, and financially care for each child. The *Family Contract* will require reasonable evidence that food, clothing, shelter, education, healthcare, and transportation will be provided to the child for a minimum duration lasting until the *Permitted* child reaches adulthood.

While the minimum duration protects the child, young adults need guidance too. The loving, respected parent or grandparent dedicated to being a good, mature role model can frequently make a significant contribution to the success of the young adults. The *Family Contract* will serve as the family's constitution, while the *Permit* will serve as the family's charter. Failure to perform under the *Family Contract* (divorce or abandonment) subjects the party to remedies for non-performance. The task of rearing children is not a casual one. Almost universally, two parents are to have greater success than one. And where even others are committed to the *Family Contract*, the incidence of success will be greater.

Questions regarding human sexuality should be dealt with within the *Family Contract*. The power of the *Libido* and its companion, the *Unconditional Response*, combine to drive, most commonly, men to promiscuous behavior. Denial of or the ignoring of the human inclination for promiscuity contributes to the very high failure rate of marriages. The possessiveness in attitudes engendered by the commitment to monogamy creates catastrophic emotional calamities when infidelity occurs. A *Family Contract* that faces the reality of promiscuity and delineates accommodations to this human trait can mitigate the catastrophic emotional stress that causes the failure of *Family Contracts* (marriages).

Human institutions must respect the nature of humanity to the extent that it contributes to a *Free-orderly society*. The *Family Contract* serves the primary purpose of producing a sound developmental environment for children and nothing more. It is not to serve the emotional well-being of the participants. Two or more people can enter into any contractual relationship with any combination of gender in a *Free-orderly society* to serve any real or imagined social, sexual, ego, or security needs, so long as it does no harm to the person, property, or minor children of others. Such is not to be confused with a *Family Contract.* At this point, many limited to conventional wisdom will see government "permitting" children to a family and a requirement of a *Family Contract* as an intolerable intrusion into the lives of citizens. Freedom is not a license to inflict harm on children and chaos in society—the inevitable result of *irresponsible procreation.* The "permitting" of children does make sense as one reviews the concepts of freedom, responsibility, and authority, applying these concepts to the issue of children and the family. Keep in mind: *Sovereign persons shall give up that sovereignty which is essential to the preservation of the remainder.*

Governing the Family in a free-orderly society

The concept of a *Free-orderly society* will be adopted by societies having vastly differing cultural traditions. So, the leadership and the process of decision-making within the family will vary widely from culture to culture and from religion to religion, even within one society. In a *Free-orderly society* that respects freedom of religion, each family is endowed with the inalienable authority to govern itself according to its own religious beliefs, values, and principles.

The *Family Contract* is between two or more sovereign adults. Each is surrendering some significant portion of its sovereignty to this partnership. It is for this reason that the terms and conditions creating this new entity should be clearly defined in a formal document.

For an orderly family in a *Free-orderly society*, a couple (yet to be married) intending to rear children should have a clear understanding and expressed agreement in the contract as to how decisions will be made and questions resolved between them. Areas of authority and responsibility must be clearly defined. The form of the *Family Contracts* will vary, being influenced by the culture, religion, and preferences of those entering into it. The *Family Contracts* will contain the minimum requirements for the permitting of a child to the family, as well as requirements of their religious group, and many provisions that will be common standard options, upon which the couple—or all—will agree. Like a team drawing a wagon, a husband and wife must be going to the same destination at the same speed and heading. Or, to say it another way, in order to succeed in the venture of raising children, they must share the same values and the same goals. Failure on this point almost assures the failure of a family. Religious leaders can assist and influence the drawing of a *Family Contract*. They can be of great assistance to their followers in defining and understanding the responsibilities and authorities they are about to assume. Where religion governs such matters, the religious leaders have an obligation to explain and obtain an expressed written commitment by the betrothed. Failure on this point is a failure to fulfill a basic role of religion. The holding of followers to high standards of behavior speaks to the worth of a religion.

There is a natural negotiation between marriage partners. Each tries to determine the limit of responsibility that can be shifted to the other, and to determine the limit of authority that can be usurped from the other. Too frequently, this can continue for years and becomes a constant source of bickering and stress to the relationship and an open sore in the family. If one partner successfully shifts responsibility and usurps authority, at some point in time, the condition becomes intolerable, and the marriage dissolves, and the family's mission to successfully rear children is threatened. A detailed *Family Contract* can minimize this vexing problem. It causes the prospective partners to face such questions and to resolve them prior to their union as a family, rather than causing rancor and bickering that impedes the family's mission of rearing children. Chaos within families fuels chaos in society. The *Family Contract* must provide for arbiters or arbitration of disputes between the partners. Arbiters will likely take many forms. The contract may provide that disputes will be resolved by religious leaders, senior family members, private agencies, among other possibilities that would serve as arbitrators—the decisions of whom the parties have agreed to abide.

The *Family Contract* then becomes the Constitution of the family, by establishing the authorities and responsibilities of each partner. It establishes the goals (direction) and the values (the path and speed) of the family's travel through life. Each family will be guided by their chosen religious values. Each must have the freedom to practice those values, whatever they are, so long as they can be practiced in a manner that does not harm the person, property, or minor children of others. The *Family Contract* can be amended by mutual consent. This is likely as the partners mature, accept, and deliberate the changes in each other as they travel through life's adventures. In a *Free-orderly society* where all citizens are sovereigns, governments will not have laws intruding into this relationship, such as marriage law.

Parental Responsibility and Authority in a Free-orderly society

The parents in a *Free-orderly society* must have total responsibility for the minor children and, in so doing, have total authority over their children. The very nature of individual freedom requires that individuals have total authority over themselves and their own progeny. There is no way for the government to encroach upon the administration of the family without encroaching on the sovereignty of the individual and the freedom to practice one's religion. In a *Free-orderly society*, there is no way for the government to usurp the authority of the family without shifting the responsibility of the family to other unwilling citizens. In a *Free-orderly society*, there is no way to assume (shift) responsibility for the family without assuming commensurate authority over the family. Thus, responsibility for minor children in a *Free-orderly society* must extend from conception to adulthood. The children obtain life at the will of the parents; their life is sustained by the efforts of the parents, and their survival to adulthood is at the will of the parents and no one else. Families undoubtedly will exercise this authority within the teachings of their religions.

Inasmuch as the parents are totally responsible for the behavior of their minor children, they must have the authority to impose any rewards or punishments they may choose as their method of behavior modification. No one can impose their judgment on others in such matters and still proclaim that they live in a *Free-orderly society*. Parents must instill inhibitions and inculcate values in their children that will produce respect for the person and property of all others. Failure to do so will result in failure as a parent, exposing the parents to substantial liability. Inasmuch as the children are the progeny of each parent, the authority over the children is vested in each parent, so the nature of the combined parental authority must be defined in the *Family Contract*. With children as with others, they cannot serve two conflicting masters. Therefore, lines of authority over the children must be sharply drawn between the parents. The *Family Contract* is virtually the only way to avoid potential conflicts and to establish the line and scope of the authority of each parent separately or jointly. This authority and responsibility must be inviolate, unless the parents are adjudicated incompetent.

Governmental intervention in family life and decisions regarding children requires the imposition of religious values. Which religion? Whose religion? In a *Free-orderly society*, only the religious values of the individual family can prevail. Religious values that may be totally acceptable to the adherents of one religion may be intolerable to those of another religion.

Children are incapable of taking responsibility for themselves and, therefore, are incapable of exercising authority over themselves. That is the very nature of childhood. In a *Free-orderly society*, at the point in time when they are capable of assuming total responsibility for themselves, as determined by law, they must be given total authority over themselves, as determined by law. Children are capable of fending for themselves at ages younger than most governments grant adulthood. For example, at sixteen, a child can drive and work, earning sufficient money to provide for the basics—food, clothing, shelter, transportation, health care, and education. This is adulthood. A sixteen-year-old adult who wants to continue schooling might enter into a contract with the parents to provide for these basic needs in consideration for accepting certain duties, standards of conduct, and levels of academic achievement, as might an eighteen-year-old. One should be reminded that the age of adulthood during the first 200 years of the United States was twenty-one. While this is a common age of adulthood in many countries, adulthood ranges between fifteen and twenty-one. Teenaged children might find these arrangements repugnant if they enjoy great authority with little or no responsibility.

In a *Free-orderly society*, no one shall sell or provide anything to a child without the consent of the parents. A child making a purchase is acting as an agent of the parents. But a child should be able to take delivery of anything for which the parent has granted authority. For example, this means that with an electronic card provided to the child by the parents, a merchant can sell to the parents and deliver to the child any services or products that are expressly approved on the card. Any other transaction must be at least a misdemeanor. The card could carry a specific monetary amount and specify permissible and impermissible transactions, such as tobacco, alcohol, drugs, sugar, pornography, or movies with ratings, or whatever is acceptable or unacceptable to the parents.

Care and Compassion in a Free-Orderly Society

The family has primary responsibility for the economic well-being, care, and compassion for all its members, the well-being of the children being foremost. It cannot be the role of government to provide care and compassion or to ensure the economic well-being of its citizens. For, as we have described, in a *Free-orderly society,* it is impossible for government to provide care and compassion without violating the property rights of those unwilling to provide care and compassion for others. Such persons might be those barely able to provide care and compassion for their own families.

Devotion and compassion of one human being for another are an innate part of human nature or Needs. These instincts have been and are essential characteristics for the survival of human life. Devotion, compassion, and love are the glue or adhesive of the family. The family is one of the institutions of society that is peculiarly able to fulfill these Needs. Religious organizations and charitable institutions have traditionally assisted the family with matters of care and compassion. As governments have undertaken these responsibilities, charities and religious groups have abandoned them.

The family, having total responsibility for itself, must have total authority over the life and death decisions that, from time to time, must be made. Among other things, this would include abortion, care of the severely disabled child, or life support for the hopelessly disabled or the dying. The family must resolve such questions within the framework of its known resources. Here again, the responsibility and authority for such decisions cannot be separated and must at all times be coextensive. A *Free-orderly society* cannot have government dictating family decisions about how to discipline children or how to resolve these vexing problems. Remember that the government cannot accept responsibility in a *Free-orderly society*, as the government can only shift responsibility and, therefore, cannot assume authority for such decisions. Families cannot expect to shift their responsibilities onto other families through the use of government and proclaim that they support a *Free-orderly society*.

Any religious group that intercedes in the decisions of its followers in matters of care and compassion must, where it exercises such authority, accept responsibility for any burdens upon society resulting from such decisions. For example, if a religious group forbids abortion among its followers, then that group must be financially responsible for whatever financial burden it causes that is beyond the ability of the family to meet the established minimum financial requirements.

Decisions as to whether or not the family must have additional children or have an abortion; decisions whether or not the family must rear hopelessly dysfunctional children or pursue infanticide; or whether or not the incurably ill must endure a painful death and sap family resources, and so forth, must remain with the family. Such decisions are difficult and must be made within the family, applying the standards, beliefs, and laws of its religion. As authority and responsibility are inseparable in a *Free-orderly society*, the family must have the responsibility and the authority for such decisions. For a religious group to assume authority in making such decisions would obligate the group to assume the responsibility attendant thereto in a *Free-orderly society*. For the government to be involved in such decisions is to usurp the family's authority and thereby enslave the family to someone else's value structure. Or for the government to assume (shift) responsibility is to enslave other families to serve such a value structure.

Because matters of care and compassion for individuals are vested in the family, it is in each individual's best interest to invest well in a family, for it is to the family they must look first for care and compassion in a time of need. Individuals who wish their assets to be applied for their personal care should make legal arrangements for this in the event of their incapacity.

The total demand for care and compassion will always exceed the resources that society can allocate to them. The supply of care and compassion provided by a society must therefore be rationed. In a *Free-orderly society*, the providers can only administer the rationing of care. Where the family is the provider, the family must evaluate its resources, both financial and human, against its priorities of responsibilities and allocate them accordingly, including nurturing of the young and care and compassion for all other family members. Adult family members requiring care must weigh their own demands on the limited resources of the family against the resources available to meet the needs for nurturing its children. As expressed under Chapter 8, religious organizations and private charitable organizations can provide backup support to families.

Professional caregivers of people of all ages, whose work results in enabling their patients to become contributors directly or indirectly to productivity and the well-being of society, make a commensurate contribution. Nonprofessional caregivers fulfill their own personal Need to provide care and compassion to others. This work is the sincerest kind of care and compassion, resulting from the very nature and Needs of human beings. It is its own reward.

The care of the hopelessly infirm, either as a result of injury, disease, or age, presents a horrible dilemma for the family. It can drain the scarce resources—time, energy, and economic assets—that would otherwise be used for the family's primary mission of rearing and nurturing its children. However difficult or painful such decisions may be, they must be made. Families, as well as all adult members of society, must be able to make such decisions. Religious congregations must assist in helping families make these decisions. It is consistent with a *Free-orderly society* for a religious group or a charity to willingly undertake to do so. As infirmities are most frequently the problem of the aged, National Fixed Investment Trusts and the ever-increasing Personal Estate Fixed Investment Trusts will cover these costs.

The burden of these decisions is so daunting that people have refused to make these decisions and have turned to the government to accept the burden. However, institutions of many descriptions have been organized to warehouse the hopelessly unproductive. We call them hospitals for the mentally ill, prisons, nursing homes, care centers, etc. The economic system described above will produce, from distributions from NFITs and inherited PEFITAs, the income necessary to support these individuals.

The Evolution of the Family in a free-orderly society

In a *Free-orderly society* that clearly defines its mission and the value of the family, the quality of the family and its product will improve infinitely. The wisdom guiding families will constantly improve, making the knowledge and understanding of all its predecessors seem primitive. In a free society, widely differing family arrangements will evolve. A highly productive father and mother may have their embryo carried by a much less productive woman, thereby freeing the more productive woman to concentrate her energies on her higher endeavor. Or perhaps the parents may procure an embryo having characteristics different from the father and mother. Family resources, which at any earlier time would seem enormous, will be allocated to the nurturing of the children. The result will be human adults with intelligence unimagined by earlier generations, as well as the disappearance of many infirmities.

A *Free-orderly society* will vastly exceed all others by having parents with the ability to have children possessing greater intelligence than their own. Through various breeding techniques in a society that has an ever-rising minimum standard for *permitting* children to a family and a continuously lengthening of healthy, productive human lives, such a society will vastly exceed all others that do not achieve these characteristics. The progressive increase of human intelligence will produce advances unimagined in traditional societies. The extension of healthy human longevity will enable experienced and wise elders to achieve their highest productivity at a time in life when those of similar age in traditional societies are feeble or dead. The *Free-orderly society* will enable each person to exploit his/her personal potential for the greatest contribution to the total advancement of society. After the passage of some time, traditional societies will, by comparison, seem backward, primitive, and pathetic.

If we accept that human life has value, then its successful continuance equally has value. The work of bearing, nurturing, rearing, and educating the subsequent generations toward having greater abilities than their predecessors is, naturally, serving that end. The primary mission of life cannot be pursued without the successful and effective procreation of the species, so that both activities are of equal value. The point reaffirmed here is that those engaged in the seemingly menial task of bearing and rearing children are engaged in the highest work of life if one subscribes to the tenet that humanity has value. In some Western societies, having no moral compass other than materialism, the rearing of children has been subordinated in importance to all sorts of other activities. Those who are not engaged in active employment but rather are engaged full-time in the tasks of providing a home and caring for children must recognize that they are doing the highest, most virtuous, and respectable work of mankind.

The Disintegration of the Families during the Twentieth Century

We know from observation and studies that children learn to pattern themselves from their parents. We know that children tend to pattern themselves after the parent of the same sex. Studies show that children patterned after the parent of the same sex are more inhibited and likely to follow accepted behavior patterns.

The twentieth-century families have disintegrated to a point where a majority of households have but one female parent. Armed with the information above, one should ask: Inasmuch as children learn by emulation and example,

"How is a male child going to learn to function as a responsible father, husband, or a male having the will and capacity for assuming responsibility, having never experienced that?"

How is a female child going to learn to function as a wife or a male child to function as a husband when neither has developed in a household, with neither having ever experienced such a relationship? The answer is that they will not and do not. The males have become feminized, and the girls have no ability to seek the strength of a responsible male, which girls learn from having a strong, loving father. With no experience or understanding of a successful marriage relationship, those who try it generally fail. Lacking a strong, responsible male figure in their development, the boys lack inhibitions—what some have called a conscience—and become dysfunctional, ending up in prison or as drug addicts.

This situation has no natural mechanism for correcting itself. Only by mandating the imposition of the standards of a *Free-orderly society* can this social cancer be cured. This is described further on. The strong family structures that largely exist among devout religious groups possess the basic qualities required of the families of a *Free-orderly society.*

The Human Potential in a Free-orderly Society

The human being is not a static creature. Regardless of how one views humanity's commencement, human beings are evolving into larger and more intelligent beings. Simple extrapolation reveals that we cannot imagine how great we may evolve in size, capabilities, intelligence, and longevity. Excluding a catastrophic event that terminates humanity, human life's capacity for perceiving the universe and economic achievement will be more profound than can now be predicted.

Within recent recorded history, human beings have grown taller and lived longer. As with other traits, tall parents tend to produce tall children. Parents having genes more conducive to greater longevity tend to pass them to their children. There exist in the human population genes for intelligence that vary by a factor of three or four. This fact alone tells us that the aggregate human intelligence is capable of advancing to vastly higher levels. There exist animal species with longevity two and three times that of humans. It is not unreasonable that the basis for that longevity might be found and adapted to the human gene pool. There exist persons with the ability to recall virtually every day of their lives, including what they were doing, what day of the week it was, and what the weather was on that day. These people are, by any other measure, quite normal. These phenomenal capacities tell us that human brain tissue has phenomenal capabilities.

One can extrapolate that "normal" human beings could someday possess such abilities.

Matter is never static, nor is human life static. While remaining mute on the question of human origins on this planet, human life will, if it were to survive in a Free-orderly society, attain longevity, abilities, and intellect beyond the comprehension of human beings at any point in time.

How society enables the procreation of those having the greatest or least abilities—social structures—can have profound effects upon the advancement and retardation of human achievement. The environment in which human life evolves accelerates or retards its advancement. This is developed at length in the concepts that follow.

All human beings are handicapped. Some are more perceptively handicapped than others. For as many capabilities as humans possess, there are human incapabilities. It is only on a current human scale of normalcy that humans are judged handicapped. On a human scale thousands of years from now, all twenty-first-century human beings will be judged incapable, handicapped, and pathetic by humans who will have lived generations in a truly Free-orderly society. To accept how handicapped the human animal is, one only needs to compare human ability to the sense of smell in lower animals, or the sight and reflexes of birds.

The organization of a society will have a substantial impact on the rate of development of the human race. Any organization of society directed to the advancement of human evolution—such as that which occurred under the coercive, totalitarian, fascist regimes of Germany and Italy in the mid-twentieth century—cannot supersede a Free-orderly society. Socialist societies that expend vast resources on the least capable pursue a path of degeneration. Those societies that adopt a Free-orderly social order will be highly conducive to ever-improving human intellect and longevity and will leave other social orders behind to muddle in ignorance and confusion. Socialist and quasi-socialist societies, whether democratic or otherwise, will fall ever farther behind a Free-orderly society to a degree comparable to that of how the twenty-first-century industrialized societies compare to the coexisting primitive societies. With the technology available even now, a tyrannous government will be able to dictate each individual's activity. Such will likely cause human society to develop into a beehive or anthill-type structure, ending human evolution.

Experience shows us that it is natural for parents to strive to produce children having greater ability and achievement than themselves. Unimpeded, parents will use all of the facilities available to achieve this end. Free families, focused on the mission of producing the next generation, will strive to produce children more capable and more productive than themselves.

CHAPTER 7

EDUCATION IN A PROSPEROUS FREE-ORDERLY SOCIETY

Education appears as a structural fabric of society that is produced by these four basic groups of institutions of civilized society. Conceived and modeled (an operating system) in the Family, it is styled (programmed) by Religion/Philosophy, sponsored by Government, and purposed and funded by the Economy. To put it in the vernacular of recent times, education is the "programming" and "uploading to memory" of the brain. It is the enabler of our marvelous computer—the human brain. Often overlooked is the fact that language is the "operating system" of our brains. The learning of language starts amazingly early, and we learn it mostly from our parents. The better the language skills of our parents, the better our "operating system. "Because of the uniqueness of our environments from conception, no two of us have the same combination of 'operating system,' 'programming,' or memories. But the better educated our parents, the greater the output and achievement of the species. Because no two of us enjoy the same brain structure, outcomes are inevitably different. Therefore, each of us must accept our uniqueness. It is because of this uniqueness applied by billions of people in billions of ways that human advancement of perception of our universe—however vast or however minute—has been achieved.

Education and Technical Training

Education is a capital asset for its holder. But it is also an asset for society. For this reason, its costs should be shared. Perhaps some scholarly endeavor can reasonably discern the proportions. In any event, the free marketplace is where it must be bought and sold. An excellent experience with such was the G.I. Bill. That was essentially a voucher system, and colleges and universities educated thousands who not only did well for themselves and their families, but also for society in general. For a society to achieve durability, it must successfully deal with this matter. There is no reason for the government to be in the school business in a society that has widely held capital assets.

Our Initial Education.

The learning process of humans begins before birth. It is the parents and the families that imprint, teach, and inculcate by example the inhibited behavior that is essential for one to function in a Free-orderly society that requires respect for the person and property of all others. Some call this one's "conscience." Studies show that this most important trait is learned exceedingly early and that identification with the parent of the same sex plays an especially important role in the development of this critical trait. The personal constraints that cause individuals to conduct themselves in a manner that respects the person and property of others are, when most successful, not by a rational process where one weighs the consequences, but rather are constraints that are a part of one's characteristics. This character of a person is molded in the person's very earliest years of life. As time goes on, an ever-greater understanding of the process that inhibits human behavior in a manner that causes one to respect the person and property of others will occur. It is during these earliest years of life when the virtues of civil, courteous behavior, honesty, the acceptance of responsibility, and respect for the person and property of others are learned and indelibly molded into the character of an individual. Achieving the ability to be self-responsible is essential for a child to become a successful adult. There is a plethora of other living habits that are taught in the home prior to the commencement of formal education. It is for this reason that a most important part of formal education must be to teach and reinforce these skills, if raising the next generation is to rank high in importance to humanity. It is valuable for those students who have had little opportunity to learn good parenting skills at home. For those who have a home that did well in teaching child-rearing skills, it helps them to hone their parenting skills. As it has been expressed here previously, if humanity has any purpose or value, the most important work of life for most human beings is the production of the next generation. The significance of this frequently eludes too many people. Women's role in this matter is so important that there are very few other activities that surpass it or are more valuable. In the family, the roles of the breadwinner and the childcare and education can be reversed. To some degree, it can also be delegated successfully. It will be rare, however, for persons of the general population to have a role to play in a society that has a purpose or mission that makes a greater contribution than that of rearing the next generation. The government, having had a major role in the destruction of the family, now exists in family structures where some forty percent of children grow up without a father in the home, or having experienced a home with a husband and a wife. The crumbled family structure has spawned the government efforts to ameliorate this problem with pre-K programs.

The Value of Formal Education

In a free society, education must be primarily the responsibility of families, religious groups, and other private organizations. All decisions regarding the education that a child receives must rest with the parents in a Free-orderly society. Any intervention in these judgments by the government is coercive and violates the sovereignty of the parents as individuals. As governmental involvement in education has taught beyond any doubt, free competition among all elements of education is essential to its efficiency and developmental progress. Tenure, unions, captive accreditation organizations, and committees for the direction of studies of the individual student have all resulted in reduced efficiency, productivity, and achievements of educational institutions. They are institutions of special interests, and they have made it costlier and more dependent upon monies from the coercive taxing power of governments. Government is not an efficient provider of education, nor can it do so free of crippling corruption, as special interests and political agendas pervert the direction and achievements of the institutions.

Each religious group has a value structure that it desires to be inculcated in the children of its families. In a Free-orderly society, each must be free to do so as long as these values include respect for persons, property, and the religious practices of all others. Government schools will fail to serve each group by either encroaching on their beliefs or by failing to respect their beliefs. Government schools are so pervasive that society must be weaned from the government breast. This can be accomplished by a system of vouchers to the parents so that as families become stronger, the subsidies will be diminished. Government schools will be sold to private school companies or placed in and operated by not-for-profit corporations. Stronger families, assisted by religious organizations and other charitable groups, and financially stronger as a result of required savings, will gradually assume responsibility for the education of their children. A Constitutional provision could precipitate this.

Irresponsible procreation has produced children lacking families with sufficient ability to provide adequate education. This has made it seem reasonable for the government to fill the void. This problem, like so many others, has as its root cause irresponsible procreation. By ending irresponsible procreation, the number of families incapable of providing education to their children will decline to a point that would make it possible for private and religious organizations to accommodate any shortfall.

Government schools have usurped the authority and responsibility of parents. In taking responsibility for the education of the children, assuming the authority naturally follows. All manner of conflicts, strife, and dissatisfaction are engendered as a result of the implementation of the government's authority. By the government assuming (shifting) the responsibility for the education of the children, it has fostered irresponsibility on the part of the parents. Funds used to pay taxes for government schools will be available to the families for tuition. Strong multi-generational families will have enhanced financial abilities to educate their children. In a free society, parents must have both the responsibility and authority over all matters relating to the education of their children.

Monopolistic government schools have stifled innovation despite the vocal efforts to the contrary. One needs only to look at the most rapidly evolving industry at the beginning of the twenty-first century to see an evolution from a government-protected monopoly of the telephone industry to an industry that provides service totally beyond the comprehension of the population only three decades earlier.

While education is a personal capital asset, a case can be made for governmental subsidy to education through vouchers. Some portion of an individual's education benefits the general well-being of society. A voucher equal to, perhaps, one-half of all education costs paid to the parents of each student with limited qualifications or encumbrances would recognize society's benefit without creating a corrupting inequality of benefits. As the principles of a durable economy are implemented, families will be more financially competent and able to accept responsibility for the education of their children. For this reason, the provision for government subsidy of education provided for in the Constitution will commence at a high rate and decline over a period of years.

Education also provides the advancement of the arts that dignify and glorify human life. They give joy to living, making the pursuit of any human mission a pleasant adventure. But in a free society, all who participate in such must do so freely and willingly, without coercion from government. Individuals who find their personal fulfillment in the arts must be free to pursue this within their means or the means of their benefactors to the extent that they do no tangible harm to others.

All successful citizens of a Free-orderly society must know what individual freedom consists of and what it entails. To be successful, all must understand the moral imperatives of a Free-orderly society. And further, all successful citizens must understand the acceptance of personal responsibility in a Free-orderly society. This is an absolutely essential function of the educational institutions of a Free-orderly society. Failure of the family to teach and inculcate, and to select schools that teach and inculcate the principles and concepts of a Free-orderly society will render their offspring dysfunctional. The early learning of civil behavior and social manners will greatly advantage each student. Verbal and gestural respect for their seniors, mentors, and elders will further advantage the students in their learning and development. Courses that make better parents will likely rank high in the curriculum for older teenagers.

To whatever extent control of government is to be vested in the citizenry, they must be educated sufficiently to perform the task adequately. Failure to understand the nature of government in a Free-orderly society and the individual's role in it is to destroy the foundation of a Free-orderly society and cause its eventual decay.

The importance of acceptance of responsibility cannot be overemphasized. Unlike the assumption of authority, the assumption of responsibility is not a natural inclination for most people. In a society for which the assumption of personal and family responsibility is paramount, the teaching and inculcation of acceptance of responsibility is an absolute necessity. Therefore, in a Free-orderly society, successful families will place the highest emphasis on producing children willing and able to accept responsibility for themselves and their eventual families. This will not only be of paramount importance within the home but will play a pivotal role in the selection of a school for their children. While these are essential elements in education, they must be adopted voluntarily by free, independent schools that understand that their freedom, independence, and survival are dependent upon their success in teaching these concepts of a Free-orderly society.

Additionally, the basic education must give each person the skills to gain employment at an entry level sufficient to provide a living wage. Schools that fail in this respect will not be worthy of the tuition paid to them by any measure.

A measure of the parents' competency is their ability and willingness to provide such education. In a Free-orderly society, parents who fail to educate their children to a level that will qualify them to be permitted parents will not become grandparents.

As schools, colleges, and universities offer a product to the public, they should be required to accurately describe what they are offering, as well as the subsequent results and outcomes. It would include their basic philosophy or religion, along with other information that will enable the parents to make a sound decision.

CHAPTER 8

RELIGION IN A PROSPEROUS FREE-ORDERLY SOCIETY

The religions of the world have provided the moral structure and the foundation of governmental law and social order. The major world religions range in age from just hundreds to several thousand years. Different people respond to different influences and the moral tenets of religions. Religious enforcement of the tenets is an important factor in maintaining an orderly society. In a Free-orderly society, religions that strengthen and add support to the family will find ever-greater followings, as more families look to their religious group and not to government to add strength to their families. Religions have provided social order, essential to human tranquility, by means of moral dogma. It is likely that some portion of the human population will always require dogmatic rules, other than those required here, to guide its behavior. Also, some portion of the human population will require the incentives of reward or punishment after death to motivate desirable behavior and inhibit undesirable behavior. As long as those behaviors fostered by each religion produce respect for the person, property, and minor children of others, each will be able to function in a Free-orderly society.

Religions offer refinements and nuances to moral behavior that are not a part of the broad, simple moral requirements of a Free-orderly society. Each serves the Needs of their followers, which are not necessarily Needs of the same magnitude as followers of other religions.

Human social life has evolved and will continue to evolve at differing rates among the various populations. At the beginning of the twenty-first century on this planet, there are societies to be found in the jungles of Borneo and the Amazon living in a way that very closely resembles the way their ancestors lived ten thousand years earlier. There are nomadic tribes living much like their ancestors of five thousand years ago. There exist many social structures having a wide range of human social development in the area of human Needs, which religion serves. As they have in the past, the practices and beliefs of religions will vary widely for the indefinite future. All must be tolerant and tolerated in a free society. The widely varying points of view and tenets held by the religions of the world will condemn societies to interminable social strife if the ability to tolerate one another's practice of their own beliefs is not strongly developed and actively assured. Each individual or family must be assured through the protection of the government that they may practice their own religion free of the values and tenets of other religions, as long as the practitioners can conduct themselves in a manner that respects the person, property, and minor children of all others. A Free-orderly society cannot tolerate religious organizations, religious leaders, or religious practitioners who expound, promote, teach, or practice harm to the person, property, or minor children of others within the society. Such is so totally and unconditionally incompatible with a Free-orderly society that it must be expunged by whatever means are available. If religious leaders are to be worthy of their position, they must strive to elevate humanity above its animalistic roots or tendencies.

As long as religions express their disagreements with other religious views within buildings or structures, and as long as they can conduct themselves in a way that respects the person, property, and minor children of others and the practice of other religions, they may thrive in a Free-orderly society. While most religions have components that seek to proselytize, in a Free-orderly society, each can do so as long as they respect the rights of others to reject their solicitations and be free of any coercion. Some religions permit infanticide and abortion of fetuses. Others do not permit abortion of fetuses but do permit killing errant minor children. Some religions permit abortion of fetuses but not infanticide, while others do not permit either. A Free-orderly society must be organized in a manner to tolerate all religions. No religion can use government to enforce its beliefs and practices on the followers of other religions or, for that matter, their own followers. The government must be disengaged and barred from involvement in matters concerning families that are governed by religious law. No matter how objectionable, horrid, or evil others may judge the family's treatment of their minor children, in a Free-orderly society, it must be tolerated to the extent that such behavior is not determined to be the work of the insane. Beyond the legal determination of insanity, no one can stand in judgment of the practices of another's religion. Each religion will thrive, languish, or perish on its merits in a Free-orderly society. As tempting as it may be, the religious leaders cannot use the coercive power of government to enforce the laws of religion on their own people. To do so would encroach upon the freedom of others to practice their religious conscience.

In Western democracies, political parties have come to be the religious organization for many who do not subscribe to traditional religion. The political parties articulate the core values for these people, providing them direction, value, and purpose. But by being a political party, it has the coercive power of government. As such, it is completely incompatible with a Free-orderly society. Having no moral code acceptable to all others, such a society is left with "Might makes Right"—the seeds for deadly combat. Some religions have punishments harsher than those that a free society might impose. In a free society, the adults of such religious groups may choose not to subject themselves to the punishments mandated by those religions. Where adulterers of a certain religion may be subject to death, the free society shall have no such penalty. However, where the choice may be between leaving one's religion or having a hand removed, the supplicant may freely choose the latter in order to stay in the grace of his/her religion. In a Free-orderly society, each religion derives its right to be practiced by virtue of the fact that all other religions have an equal right to be practiced. Families must have an unimpeded right to practice their religion and to manage, discipline, provide, or withhold care to or from their minor children or adult members who have expressly submitted to the judgment of the other adult members. Any intrusion into the conduct of the family, beyond the initial Permitting of children, would require the imposition of religious values differing from those held by the family and would be incompatible with a Free-orderly society. As described previously, the religious practice of unlimited birth or irresponsible procreation cannot be accomplished without adversely impacting others in society. So, in the context of a Free-orderly society, religions must yield on this point. Unwillingness to do so would require emigration to a country that permits such behavior.

To remind the reader again, some would dismiss these concepts as being Utopian or idealistic and therefore unattainable. This, however, implicitly acknowledges that these concepts point the direction in which societies should move. We acknowledge, however, that contemporary societies are some distance from practicing these concepts, even those that think of themselves as being free. Because these concepts appear to some as unattainable does not mean none is attainable. And appearing unattainable does not render them invalid.

Strengthening Families

To the extent that religious groups are families of families joined together to strengthen each other, religions are challenged by the evolving social and economic structures. Mobility causes families to lose their roots, leaving grandparents, aunts, uncles, and friends far behind. Religious congregations can fulfill a role of supporting and nurturing children if they assist these "abandoned" relatives and friends to become attached to and serve an active role as part of another family. Those families that have moved off may seek others who have been "abandoned" to become appended and supportive of their family. If we can accept that rearing the next generation is among the highest works of life, then such relationships provide lives with real value. The emotional rewards vastly exceed those of playing another game of golf, tennis, bridge, or any number of activities that use up time while producing no tangible results.

The Christian concept of godparents could do much to fulfill voids created by the loss of a parent if the godparents were truly involved and serving an active role in the family. Religions can contribute much by promoting the value of childless adults becoming appended to families who have lost a parent, or assisting grandparents who are removed from their own grandchildren to become appended to and serve an active role in families that do not have active grandparents. It is hard to imagine—and probably harder to find—a child who has too much competent and supportive parenting and love.

Religious congregations and their greater organizations can combine families into economic units in order to improve buying power in the marketplace. Religions can undertake insurance functions, providing their member families with education assurance, health insurance, income annuities, and insurance to cover losses caused by death and casualty, among other functions. There are long traditions of religious organizations providing such functions. These are manifested as religious schools and universities, religion-sponsored hospitals, religion-sponsored charities, and religious congregational assistance in replacing the destroyed homes and barns of its members.

Inculcate Civility

Civility distinguishes the human animal from all others. It is the ability to conduct oneself at all times in a manner that respects the person and property of others. Considerate treatment of others, thoughtful conduct around others, and the practice of good etiquette are the lubricants of social interaction. The religion of a Free-orderly society must teach and demand civility among its followers, who in turn must reject, repudiate, and in general be intolerant of crude, crass, and uncivil behavior. Traditional religions all possess the teachings of civil behavior, but, frequently, their tolerance of transgressors has bred socially debilitating, uncivil behavior.

Somewhere in the human psyche is a morbid fascination with aggression, combat, mayhem, and strife. This manifests itself in electronic games, the cinema, athletic games, automobile racing, bullfighting, and, in an earlier time, the gladiatorial combat of Rome—not to mention war itself. Those who concern themselves with matters of civility may find fertile ground for the understanding of this human propensity.

Strengthen the Individual

Each individual lives with a multiplicity of internal conflicts. These arise from conflicting personal Needs or motivations. Also, they arise from emotional, psychological, or physical conflicts within the individual. The challenge of religious groups in a Free-orderly society is to assist the individual directly or indirectly, through the sponsoring of institutions, to cope with and manage these personal internal conflicts.

A conflict that is particularly detrimental to the soundness of the family is the general incompatibility of the male sexual drive with the monogamy required of a successful male role in the family. The inclination or drive toward promiscuity, predominantly a male problem, needs to be recognized and dealt with in a manner that will enable individuals to successfully fulfill their parental roles in the family. The traditional method of dealing with this conflict, through the denial of this human nature, has been a pathetic failure. The challenge to religious groups is to recognize this human nature and to devise or accept institutions that solve the personal conflicts, enabling the fulfillment of responsibility and commitments to the families. It may be in the form of preparing the newlyweds, who take traditional vows of abstinence and fidelity, for their future frustrations resulting from the denial of human desires, and that they must subordinate their own desires to the obligation of rearing children.

Another conflict of human nature arises from the human motivations of jealousy, envy, and covetousness. In a free society, great disparities in personal or family wealth and income develop. These disparities result from natural differences in genetic endowment and the capriciousness of good or bad fortune. People who wish to live in a free society must be able to tolerate these disparities. These instincts are deeply embedded in the human psyche and manifest themselves in disruptive behavior. These emotions also commonly accompany the emotions of love, producing another challenge to those who would provide religious and personal counseling. It will be the challenge of religious groups to enable followers, as a group, to manage these very human emotions of jealousy, envy, and covetousness. Unfortunately, the new political religions appear to traffic in these base human instincts.

Human virtues

Human virtues include the whole range of human endeavor that has raised humanity to higher levels of civility and accomplishment. Personal responsibility for one's and one's family is an essential virtue of all who live in a Free-orderly society. Lessons promoting the growth of this important virtue are a prime work of any religious teaching. Self-discipline is a virtue that is both fostered and essential in a Free-orderly society. This virtue affects the total life of the individual, from one's productivity to the whole range of matters that contribute to one's personal fulfillment.

Also, a challenge for religions is to inculcate the virtue of self-denial. For those persons who wish to rear children, such an undertaking requires parents to be able to deny themselves for the well-being of the children. This will take the form of the parents denying themselves material things, free time, personal fulfillment, and sexual gratification that would interfere with family achievement. The preparation for one's security in the event of disability from accident, disease, or age requires the ability to deny oneself immediate gratification and to save and insure for such events. Self-denial in the form of forbearance of current consumption is the essential basis for the accumulation of wealth. It is this process that reduces the disparity of wealth in the population.

The virtue of personal integrity is essential to a Free-orderly society. Honesty, measured by the requirement that people must behave in a manner that respects the person and property of all others, is an essential virtue. By that measure, dishonesty that harms no one is of no consequence. Unless it can be demonstrated to the contrary, telling children that there is a Santa Claus is a dishonesty that, evidently, over the years, has harmed no one. Far too frequently, dishonesty, intentionally or unintentionally, harms someone's person, property, and/or minor children.

Reliability is an essential virtue. When a person commits to perform a task, to pay a sum, or to accept a responsibility, fulfillment is an essential virtue. Loyalty to one's employer and loyalty to one's employees are essential virtues.

The loyal employee is the extension of the employer's arm—his agent acting on his employer's behalf. Reliability on the part of the employer to fulfill his commitment to the employee is equally an essential virtue.

Having pride in one's work is an essential virtue. It not only assists the success of one's personal endeavor, but it also assists the success of the greater project that improves the lot of all. It also gratifies the human Need, ego, an important factor in personal happiness.

Service in government, with the dedication to preserving a Free-orderly society, is a high virtue that should be aspired to by the most experienced and most qualified. It should be given without consideration for remuneration, and those who serve should receive accolades and respect, recognizing their high achievements and service. Service in defense of a Free-orderly society is commonly at risk to one's life, and such service is of the highest virtue. A Free-orderly society must grant its highest respect to those who serve in this capacity. These are but a few of the virtues that religious teachings confer on their followers.

PERCEPTIONISM® is a philosophy that possesses some characteristics of a religion. It offers humanity purpose, value, and direction in The Human Mission.

PART THREE

OFFERS AN OUTLINE FOR A CONSTITUTION WHICH WOULD ENABLE A FEE-ORDERLY SOCIETY.

CHAPTER 9

A CONSTITUTION FOR A PROSPEROUS FREE-ORDERLY SOCIETY

A practical application of the concepts of a Durable Growing Economy and a Free-orderly society to a greatly modified Constitution of the United States of America

A modification or restyling of the Constitution of the United States of America that conforms to the requirements of a

Free-orderly Society is offered here as an outline for all who seek to form a constitutional government for a Durable Growing Economy and a Free-orderly Society. The Constitution of the USA was a major step in the direction of a republican form of government, and the experiences of over two hundred years have produced enlightenment about which elements contribute to a Prosperous, Free-orderly society and which do not.

When one examines the Constitution of the United States of America and applies the principles of a Free-orderly Society and the requirements of a Durable Growing Economy, some very interesting modifications to the document emerge. Quite incredibly, much of the original document and the intent of the original document embody the principles of a Free-orderly Society. But when applying the principles of a Free-orderly Society to the Constitution of the USA, one finds that much of it must be put back the way it was originally, with, of course, the exception of those amendments that corrected the matters that failed to treat all human beings equally under the Constitution. Some additions must be made to embody the original intent of the founders and to abolish the abuses that the founders did not intend or anticipate. For those whose vested interest will cause them to resist the changes in the structure of the government of the USA, this language of the founders in the Declaration of Independence might be a sound admonishment:

"We hold these truths to be self-evident, that all men are created equal, that they are endowed by their Creator with certain unalienable Rights, that among these are Life, Liberty and the pursuit of Happiness.—That to secure these rights, Governments are instituted among Men, deriving their just powers from the consent of the governed.—That whenever any Form of Government becomes destructive of these ends, it is the Right of the People to alter or to abolish it, and to institute new Government, laying its foundation on such principles and organizing its powers in such a form, as to them shall seem most likely to affect their Safety and Happiness."

Provisions may be scheduled for delayed implementation for various reasons. One obvious reason will be for those provisions that adversely affect deciding officials—they can be delayed to a time when these officials will no longer be affected.

Whatever country were to adopt this Constitution, it will become a magnet for the most productive people on Earth. Its economic, social, and political structure will prove so superior to all others that it will be copied.

Embodied in this Constitution are many elements that one might think more appropriately just matters for legislation. Unfortunately, human nature being what it is, many matters are too important to be left to the domain of politicians to be easily changed. The importance and monetary significance are so great as to be able to corrupt most people.

All must be aware that the adoption of a document such as this will require an evolution of social and institutional norms that may take decades before it can be fully implemented and memorialized.

The Assembly

Continuing our metaphor, we will assemble the many parts into a sleek, smooth-running vehicle that possesses the software that will keep it traveling on the road, at a safe speed, with an infinite fuel supply. For a rational destination, we offer THE HUMAN MISSION.

It is evident that for a society to transcend from one that has rendered a large portion of its population incapable of caring for itself to one of predominantly individuals capable of caring for and being responsible for themselves, it will require a period of time for the transition. For this reason, there is a provision for a temporary "economic safety net." Those who will style a Constitution for their society based upon this outline may well need to prescribe periods and graduated methods for implementation in order to accommodate the transition.

The following is taken from the original Constitution of the United States of America, most of which has served effectively for over two hundred years. Attempted here is a correction of the oversights, perversions through amendments, and the corruptions by the political and legal/judicial systems.

To many, that which appears here might be better left to lawmakers. Experience teaches us that whatever discretionary power people have, it will be exercised in a manner that serves their best interests. For this reason, discretionary power must be limited wherever and whenever possible. Explanations will occasionally appear and will be in italics. It is based upon the premise that this outlined Constitution is to produce, provide, preserve, and protect a Free-orderly society possessing a Durable Growing Economic structure.

The Constitution of Country/Union:

We, the Individually Sovereign People, who accept total authority and responsibility for our Person, Property, and Minor children, relinquish only that sovereignty necessary to subscribe to the provisions of this document in order to form a Free-orderly Society having a Durable Growing Economy and grant the establishment of a Government having the SOLE intent of Producing, Providing, Preserving, and Protecting such a Society. The Freedom and Sovereignty of each Adult Citizen shall be SUPERIOR to Human Life itself, and the Needs of human life shall be SUPERIOR to all other life forms when administering or interpreting this Constitution. In so doing, the Government shall produce and provide only those institutions necessary to achieve these ends. These will include laws assuring: the establishment only of sound family structures; enforcing legal contracts; ensuring competitive, transparent markets free of fraud and deception; providing for the institutions of commerce; for the institutions of social intercourse; mandating and enforcing civilly responsible behaviors among and between sovereign persons; guaranteeing equal application of the laws to all of the people without exception; domestic tranquility; and providing for effective protection of the people and the defense of the (Country/Union). We, the Sovereign People, ordain and establish this Constitution, which shall govern all people who choose to and are permitted to reside in (Country/Union).

Beyond the establishment of the Family, the government shall have no additional authority in its governance. The government will not be involved directly in any matters of "care and compassion," beyond providing a basic economic safety net. That will eventually be left to the sovereign citizens and eleemosynary institutions.

ARTICLE I. LEGISLATIVE POWERS

Section 1

All legislative Powers herein granted shall be vested in a Congress of (Country/Union), which shall consist of a Senate and House of Representatives.

Section 2

The House of Representatives shall be composed of Members chosen every six years by the Sovereign People on the basis of one Adult Elector from each Household, irrespective of Gender. The required attained age for Adulthood shall be prescribed by law.

One must keep in mind that freedom and a Free-orderly society are neutral, and the Household/Family is the basic Social, Educational, economic, Psychological, and Spiritual Institution. It is this institution that is the foundation of society, that, as such, must choose its elector or voter. Remember, the purpose of the Government is not to represent any special group, special interests, or gender – its only purpose is to produce, provide, preserve, and protect a Free-orderly Society having a Durable Growing economic structure for all of the households/families, and as such, the interest of one is essentially the interest of all. Any other interests would not be compatible with a Free-orderly society.

Section 3

No Person shall be a Representative for more than one six-year term in a lifetime, commencing with the next term of Congress, or who shall not have attained the Age of fifty Years, and who shall not, when elected, have been a Resident for the preceding five years of the District which he/she shall be chosen, or who has ever been an attorney-at-law. Once having served, an ex-Representative shall be disqualified from serving as an attorney-at-law for ten years after serving as a Representative.

Time has taught that power corrupts, and the only way to avoid such corruption is to limit that power and to limit the terms of power. The self-serving, professional politician destroyed the system of government intended by Thomas Jefferson and James Madison for the United States of America. They have done this by placing the determination of their continuance in office in the hands of the least competent and most easily swayed and corrupted. To correct this condition, the Constitution must require the establishment of term limitations on all officeholders. Wherever and whenever such provisions must be implemented by sitting officials, a delayed implementation period of perhaps thirty years will be required to obtain the required approval.

Attorneys-at-law have successfully exploited the legislative process for professional advantage since the first century in Rome, so that an attorney-free House of Representatives might be able to prevent the profession from creating an advantage for itself at the expense of others.

Section 4

No group, coalition, association, or political party may move or conspire to control Electors or influence the election of any Officeholder, or the legislation or decisions of any Officeholder or government official in any way not provided for in this Constitution.

The Congress shall establish punishments that effectively deter any such activity.

Political parties have proven themselves to be coalitions that conspire to use the coercive power of government to benefit themselves at the expense of others. This is the antithesis of a free society, as coercion is the antithesis of freedom, and in this section, they are prohibited.

Section 5

They shall be divided as equally as may be into three Classes. The Seats of the Representatives of the first Class shall be vacated at the expiration of the second year, of the second Class at the expiration of the fourth year, and of the third Class at the expiration of the sixth year, so that one-third may be chosen every second year.

This is in order to provide continuity to the body.

Section 6

Certification of candidates for election to the House of Representatives will require the endorsement by at least four organizations, companies, religious groups, or governments in the District that collectively represent citizens, residents of the district, equal to one-fourth the total adult population of the district. No election effort may expend a sum, from any source, greater than the equivalent of one dollar, adjusted for inflation of the currency, per adult citizen of the District.

This is intended to minimize the efforts of some to place in the House of Representatives persons who will do their bidding or blanket enforcements that lack critical evaluation.

Section 7

The Number of Representatives shall not exceed 500, but each (State/Province/Nation) shall have at least one Representative. When vacancies happen in the Representation from any

(State/Province/Nation), the Executive Authority thereof shall issue Writs of Election to fill such Vacancies. The House of Representatives shall choose its Speaker and other Officers; and shall have the sole Power of Impeachment.

Section 8

The Senate of *(Country/Union)* shall be composed of two Senators from each (State/Province/Nation) having five or more Representatives, or one Senator from each (State/Province/Nation) having fewer than five Representatives, chosen by the Legislature thereof for six Years; and each Senator shall have one Vote. No Person shall be a Senator for more than one term in a lifetime, or who shall not have attained the Age of fifty-five years, or who has not been a resident of the (State/Province/Nation) for which he/she shall be chosen for ten years. Each (State/Province/Nation) Senate shall nominate five persons to serve as Senator, from whom the House of Representatives of that (State/Province/Nation) will elect a Senator. Only one Senator from each state may have been an Attorney-at-law.

Experience and maturity are essential qualities of good judgment. Such persons will likely have achieved a successful career and be able to serve as statesmen. Subjecting the Senatorial officeholders to popular election has proven to be a debilitating and corrupting process.

They shall be divided as equally as may be into three Classes. The Seats of the Senators of the first Class shall be vacated at the expiration of the second year, of the second Class at the expiration of the fourth year, and of the third Class at the expiration of the sixth year, so that one-third may be chosen every second year; and if vacancies happen by Resignation or otherwise during the Recess of the Legislature of any (State/Province/Nation), the Executive thereof may make temporary Appointments until the next Meeting of the Legislature, which shall then fill such Vacancies.

This is to provide continuity to the body of the Senate.

Section 9

The Vice President of the *(Country/Union)* shall be President of the Senate, but shall have no Vote, unless they be equally divided.

Section 10

The Senate shall choose its other Officers, and also a President pro tempore, in the absence of the Vice President, or when he shall exercise the Office of President of *(Country/Union)*.

Section 11

The Senate shall have the sole Power to try all Impeachments. When sitting for that Purpose, they shall be on Oath or Affirmation. When the President of *(Country/Union)* is tried, the Chief Justice shall preside; and no Person shall be convicted without the Concurrence of two-thirds of the Members present. Judgment in Cases of Impeachment shall not extend further than to removal from Office, and disqualification to hold and enjoy any Office of respect, Trust, or Profit; but the Party convicted shall nevertheless be liable and subject to Indictment, Trial, Judgment, and Punishment, according to Law.

Section 12

The Times, Places, and Manner of holding Elections for Representatives shall be prescribed in each (State/Province/Nation) by the Legislature thereof; but the Congress may at any time by Law make or alter such Regulations, except as to the Places of choosing Senators.

The Congress shall assemble at least once in every year, and such meeting shall be on the ***Date***, unless they shall by Law appoint a different Day.

Section 13

Each House shall be the Judge of the Elections, Returns, and Qualifications of its own Members, and a Majority of each shall constitute a Quorum to do Business; but a smaller number may adjourn from day to day, and may be authorized to compel the Attendance of absent Members, in such manner, and under such Penalties as each House may provide. Each House may determine the Rules of its Proceedings, punish its Members for disorderly behavior, and, with the concurrence of two-thirds, expel a Member. Each House shall keep a Journal of its Proceedings, and from time to time publish the same, excepting such Parts as may in their Judgment require Secrecy; and the Yeas and Nays of the Members of either House on any question shall, at the desire of one-fifth of those present, be entered on the Journal. Neither House, during the Session of Congress, shall, without the consent of the other, adjourn for more than three days, nor to any other place than that in which the two Houses shall be sitting.

Section 14

The Senators and Representatives shall serve for the Honor and receive no compensation other than for their expenses related to fulfilling their duties and responsibilities of the office, to be paid out of the Treasury of *(Country/Union)*. They shall, in all Cases except Treason, Felony, and Breach of the Peace, be privileged from Arrest during their Attendance at the Session of their respective Houses, and in going to and returning from the same; and for any Speech or Debate in either House, they shall not be questioned in any other Place.

Section 15

No Senator or Representative shall, during the Time for which he was appointed or elected, be appointed to any civil Office under the Authority of (Country/Union), which shall have been created, or the Emoluments whereof shall have been increased during such time; and no Person holding any Other Public/Government Office shall be a Member of either House during his Continuance in Office. Other than the aforementioned, no Member of Congress shall enjoy exemption from or benefits greater than the laws applicable to the remainder of the population, except for reimbursed expenses.

Experience has demonstrated the propensity of legislators to provide themselves with exemptions and benefits separate from the remainder of the population.

Section 16

No person shall be a Senator or Representative in Congress, or elector of President and Vice President, or hold any office, civil or military, under the (Country/Union), or under any (State/Province/Nation), who, having previously taken an oath as a Member of Congress, or as an officer of the (Country/Union), or as a Member of any (State/Provincial/National) Legislature, or as an executive or judicial officer of any (State/Province/Nation), to support the Constitution of the (Country/Union), shall have engaged in insurrection or rebellion against the same, or given aid or comfort to the enemies thereof. But Congress may, by a vote of two-thirds of each House, remove such disability.

Section 17

All Bills for raising Revenue shall originate in the House of Representatives; but the Senate may propose or concur with Amendments as on other Bills. Every Bill which shall have passed the House of Representatives and the Senate shall, before it becomes a Law, be presented to the President of the (Country/Union). If he approves, he shall sign it; but if not, he shall return it, with his Objections, to that House in which it shall have originated, who shall enter the Objections at large on their Journal and proceed to reconsider it. If, after such Reconsideration, two-thirds of that House shall agree to pass the Bill, it shall be sent, together with the Objections, to the other House, by which it shall likewise be reconsidered, and if approved by two-thirds of that House, it shall become a Law. But in all such Cases the votes of both Houses shall be determined by Yeas and Nays, and the Names of the Persons voting for and against the Bill shall be entered on the Journal of each House respectively. If any Bill shall not be returned by the President within twelve Days after it shall have been presented to him, the same shall be a Law, in like manner as if he had signed it, unless the Congress by their Adjournment prevents its Return, in which Case it shall not be a Law.

Every Order, Resolution, or Vote to which the Concurrence of the Senate and House of Representatives may be necessary (except on a question of Adjournment) shall be presented to the President of (Country/Union); and before the Same shall take Effect, shall be approved by him, or being disapproved by him, shall be repassed by two-thirds of the Senate and House of Representatives, according to the Rules and Limitations prescribed in the Case of a Bill.

Section 18

No Bill or Act shall contain matters or provisions not specifically pertaining to the subject Act or Bill.

This is to avoid the abuses that developed in the USA, called "earmarks" and such.

Section 19

No Bill or Act shall contain matters or provisions that preserve or extend life in any form or any individual life that is not essential to the continuance of this Free-orderly Society.

Section 20

All who serve in the Congress shall take the following oath of office:

"I solemnly pledge to fulfill the duties of this office, committing the best of my abilities to honor that which is set forth in the Constitution of (Country/Union)."

Article II. Executive Power

Section 1

The executive Power shall be vested in the President of the (Country/Union). He/She shall hold his/her Office during a single Term of Six Years in a lifetime, and, together with the Vice President, chosen for the same Term, be elected as follows: Each (State/Province/Nation) shall appoint, in such manner as the Legislature thereof may direct, a number of Electors, equal to the whole number of Senators and Representatives to which the (State/Province/Nation) may be entitled in the Congress; but no Senator or Representative, or Person holding an Office of Trust or Profit under the (Country/Union), shall be appointed an Elector. Except for conditions provided herein, all Electors shall serve absolutely unencumbered. No group, coalition, association, or political party may move or conspire to control Electors or influence the election of any Officeholder in any way not provided for in this Constitution. The Congress shall establish punishments that effectively deter any such activity.

The inevitable deriding of a President standing for re-election with respect to foreign policy is tantamount to deriding the government of the country. This weakens the presidency and, thereby, the government as it deals with other nations. It must be kept in mind that for the government of a Free-orderly Society, foreign policy and dealing with other nations is the paramount responsibility of the President. Further, this would end the compromising of the popular election processes and the vast waste of resources on the election, and the deleterious distractions that dominate the attention of the office of the President.

Political parties have proven themselves to be coalitions that conspire to use the coercive power of government to benefit themselves at the expense of others. This is the antithesis of a free society, as coercion is the antithesis of freedom, and in this section, they are prohibited.

The Electors shall meet in their respective (States/Provinces/Nations), and after making this solemn pledge:

"I vow that to the best of my ability to select a person who complies with the requirements of this Constitution and that I believe will serve the intent of it,"

Will vote by Ballot for two Persons from a list of ten, five nominated by the House of Representatives and five nominated by the Senate, of whom at least one shall not be an Inhabitant of the same (State/Province/Nation) with themselves. And they shall make a list of all the Persons voted for, and of the number of Votes for each; which List they shall sign and certify, and transmit sealed to the Seat of the Government of the (Country/Union), directed to the President of the Senate.

The President of the Senate shall, in the Presence of the Senate and House of Representatives, open all the Certificates, and the Votes shall then be counted. The Person having the greatest Number of Votes shall be the President, if such Number be a Majority of the whole Number of Electors appointed; and if there be more than one who have such Majority and have an equal Number of Votes, then the House of Representatives shall immediately choose by Ballot one of them for President; and if no Person have a Majority, then from the five highest on the List, the said House shall in like Manner choose the President. But in choosing the President, the Votes shall be taken by (States/Provinces/Nations), the Representation from each (State/Province/Nation) having one Vote; a quorum for this purpose shall consist of a Member or Members from two-thirds of the (States/Provinces/Nations), and a Majority of all the (States/Provinces/Nations) shall be necessary to a Choice.

In every Case, after the Choice of the President, the Person having the greatest Number of Votes of the Electors shall be the Vice President. But if there should remain two or more who have equal Votes, the Senate shall choose from them by Ballot the Vice President.

The Congress may determine the Time of choosing the Electors, and the Day on which they shall give their Votes; which Day shall be the same throughout the (Country/Union).

No Person except one born of two Citizens shall be eligible to the Office of President or Vice President; neither shall any Person be eligible to those Offices who shall not have attained the Age of Fifty-five Years. In case of the removal of the President from office, or of his death or resignation, the Vice President shall become President.

Section 2

Whenever there is a vacancy in the office of the Vice President, the President shall nominate a Vice President, conforming to the above requirements, who shall take office upon confirmation by a majority vote of both Houses of Congress.

Section 3

Whenever the President transmits to the President pro tempore of the Senate and the Speaker of the House of Representatives his written declaration that he is unable to discharge the powers and duties of his office, and until he transmits to them a written declaration to the contrary, such powers and duties shall be discharged by the Vice President as Acting President.

Section 4

Whenever the Vice President and a majority of either the principal officers of the executive departments, or of such other body as Congress may by law provide, transmit to the President pro tempore of the Senate and the Speaker of the House of Representatives their written declaration that the President is unable to discharge the powers and duties of his office, the Vice President shall immediately assume the powers and duties of the office as Acting President.

Thereafter, when the President transmits to the President pro tempore of the Senate and the Speaker of the House of Representatives his written declaration that no inability exists, he shall resume the powers and duties of his office unless the Vice President and a majority of either the principal officers of the executive departments, or of such other body as Congress may by law provide, transmit within four days to the President pro tempore of the Senate and the Speaker of the House of Representatives their written declaration that the President is unable to discharge the powers and duties of his office. Thereupon, Congress shall decide the issue, assembling within forty-eight hours for that purpose if not in session. If the Congress, within twenty-one days after receipt of the latter written declaration, or, if Congress is not in session, within twenty-one days after Congress is required to assemble, determines by two-thirds vote of both Houses that the President is unable to discharge the powers and duties of his office, the Vice President shall continue to discharge the same as Acting President; otherwise, the President shall resume the powers and duties of his office.

Section 5

The terms of the President and Vice President shall end at noon on the twentieth day of January, and the terms of Senators and Representatives at noon on the third day of January. The Congress may by law provide for the case of the death of any of the persons from whom the House of Representatives may choose a President whenever the right of choice shall have devolved upon them, and for the case of the death of any of the persons from whom the Senate may choose a Vice President whenever the right of choice shall have devolved.

Section 6

The President shall, at stated times, receive for his services a compensation, which shall neither be increased nor diminished during the period for which he shall have been elected, and he shall not receive within that period any other emolument from the (Country/Union).

Section 7

The President shall be Commander in Chief of the Armed Forces of the (Country/Union), and of the Militia of the several (States/Provinces/Nations) when called into the actual Service of the (Country/Union); he shall delegate all matters of defense of the (Country/Union) to the military, the plans and actions of which will be confined to the defense and survival of the (Country/Union), and will be subject to endorsement or rejection by the Congress; he may require the opinion, in writing, of the principal officer in each of the executive departments, upon any subject relating to the duties of their respective offices.

Section 8

He/She shall have power, by and with the advice and consent of the Senate, to make treaties, provided two-thirds of the Senators present concur. With the exception of treaties dealing with Earth's indivisible resources, no treaty shall be SUPERIOR to this Constitution, and no provisions of this Constitution shall be subordinated to or be abridged by any treaty. The President shall nominate, and by and with the advice and consent of the Senate, shall appoint Ambassadors, other public Ministers and Consuls, Judges of the Supreme Court, and all other Officers of the (Country/Union), whose appointments are not herein otherwise provided for, and which shall be established by law; but the Congress may by law vest the appointment of such inferior officers, as they think proper, in the President alone, in the Courts of Law, or in the Heads of Departments. All such appointed officers and officials shall take the oath:

"I solemnly pledge to execute the duties of my office in a manner that, as provided in the Constitution, protects the sovereignty of each person that falls within the purview of my decisions."

Section 9

The President or those to whom he has delegated authority shall neither make nor enforce decisions that preserve life in any form that is not essential to the preservation of the Free-orderly Society.

Section 10

The President shall from time to time give to the Congress information of the state of the (Country/Union), and recommend to their consideration such measures as he shall judge necessary and expedient; he may, on extraordinary occasions, convene both Houses, or either of them, and in case of disagreement between them, with respect to the time of adjournment, he may adjourn them to such time as he shall think proper; he shall receive Ambassadors and other public Ministers; he shall take care that the laws be faithfully executed, and shall commission all the officers of the (Country/Union).

Section 11

The President, Vice President, and all civil officers of the (Country/Union) shall be removed from office on impeachment for, and conviction of, treason, bribery, or other high crimes and misdemeanors.

Section 12

The President and the Vice President shall take the following oath prior to undertaking the duties of their office:

"I solemnly pledge to fulfill the duties of this office to the best of my ability to preserve, protect, and defend the Constitution of (Country/Union)."

Article III. Judicial Power and Courts

Section 1

The judicial power of the (Country/Union) shall be vested in one Supreme Court, consisting of nine Justices, and in such inferior Courts as the Congress may from time to time ordain and establish. The Judges, both of the Supreme and inferior Courts, shall hold their offices during good behavior not to exceed their eightieth year of age, and shall, at stated times, receive for their services a compensation, which shall not be diminished during their lifetimes.

Section 2

The judicial power shall extend to all cases, in law and equity, arising under this Constitution, the laws of the (Country/Union), and treaties made, or which shall be made, under their authority;—to all cases affecting Ambassadors, other public Ministers and Consuls;—to all cases of admiralty and maritime jurisdiction;—to controversies to which the (Country/Union) shall be a party;—to controversies between two or more (States/Provinces/Nations);—between citizens of different (States/Provinces/Nations);—between citizens of the same (State/Province/Nation) claiming lands under grants of different (States/Provinces/Nations); and between a

(State/Province/Nation), or the citizens thereof, and foreign States, Citizens or Subjects.

The judicial power of the (Country/Union) shall not be construed to extend to any suit in law or equity, commenced or prosecuted against one of the (States/Provinces/Nations) by citizens of another (State/Province/Nation), or by citizens or subjects of any foreign state.

Section 3

In all Cases affecting Ambassadors, other public Ministers and Consuls, and those in which a (State/Province/Nation) shall be a Party, the Supreme Court shall have original Jurisdiction. In all the other Cases before mentioned, the Supreme Court shall have appellate Jurisdiction, both as to Law and Fact, with such Exceptions, and under such Regulations as Congress shall make.

The Trial of all Crimes, except in Cases of Impeachment, shall be by Jury; and such Trial shall be held in the (State/Province/Nation) where the said Crimes shall have been committed; but when not committed within any (State/Province/Nation), the Trial shall be at such Place or Places as the Congress may by Law have directed.

Section 4

Treason against the (Country/Union) shall consist only in levying War against them, or in adhering to their Enemies, giving them Aid and Comfort. No Person shall be convicted of Treason unless on the Testimony of two Witnesses to the same overt Act, or on Confession in open Court. The Congress shall have the Power to declare the Punishment of Treason, but no Attainder of Treason shall work Corruption of Blood or Forfeiture.

Section 5

All Judges shall take this Oath upon accepting office – "I do solemnly swear (or affirm) that I will faithfully preserve and protect the integrity of the Constitution of the (Country/Union) and all of my judgments, determinations, and decisions shall be based exclusively on the Constitution."

Section 6

Attorneys at Law practicing in these courts shall not participate directly or indirectly in judgments of monetary compensation awarded to the complainant beyond customary fees and expenses. Any and all punitive damages awarded shall become the property of the Government.

Contingent fees for attorneys produced a cripplingly litigious society. Extensive litigation in a society is like plaque in the arteries. Punitive damages are, in effect, a fine or penalty exacted to modify undesirable behavior, having nothing to do with the complainant and therefore belong to the government.

Section 7

All laws will be judged on the basis that if they encroach upon or diminish the Sovereignty of the individual, such encroachment or diminishment is deemed essential to the preservation of the remaining individual's Sovereignty over their Person, Property, and Minor Children.

Section 8

If any case before any court exceeds three years without adjudication, the jurists may be dismissed by a majority vote of the Senate Judiciary Committee.

Section 9

Congress shall determine a minimum education level for all Jurors, and Jurors shall be compensated a sum that collectively equals six times the working daily compensation paid to the presiding judge.

Section 10

Congress shall have the Power to declare the Punishment of False Witness or accusation in the Courts, but the punishment must be an effective deterrent and punishment imposed as written.

Article IV. Law Enforcement Powers

Section 1

All investigations of wrongdoing and enforcement of the law shall be vested in Grand Juries consisting of seven Jurists to be appointed as follows: The President shall appoint two Directors; the Justices shall appoint two Directors; the Senate shall appoint one Director; and the House of Representatives shall appoint two Directors.

Initial appointments shall be such that three from separate branches shall be for terms of two years; two from separate branches for four years; and two from a separate branch for six years. Thereafter, all appointments shall be for one six-year term.

Section 2

They shall be divided as equally as may be into three Classes. The Seats of the Representatives of the first Class shall be vacated at the expiration of the second year, of the second Class at the expiration of the fourth year, and of the third Class at the expiration of the sixth year, so that one-third may be chosen every second year.

Section 3

They shall designate an Attorney-General and a Chief Investigator. If reasonable cause for prosecution is found, the matter will be presented to the Courts.

Section 4

Rights to non-self-incrimination must be honored. Neither information derived from coercive invasion of one's organic memory nor the invasion of one's electronic memory shall be admissible in Court.

Article V. Government Powers and Limitations

Section 1

The Government shall have Power:

To lay and collect Consumption Taxes, Duties, Imposts, Fees, and Excises, to pay the Debts and provide for the common defense and cost of administration of the (Country/Union); but all Consumption Taxes, Duties, Imposts, Fees, and Excises shall be assessed uniformly to all consumer products and services without exception and be uniform throughout the (Country/Union).

It may impose a tax on those who do not rear sufficient children to maintain society and distribute the proceeds without discrimination to those who rear more than the amount required to maintain society.

Government officials can have no discretion over the levying of consumption taxes other than the total amount on all products and services or allocating proceeds, so as to avoid the inevitable corruption.

If, for example, it is determined that it will require each family to rear two and six-tenths children to maintain the society, each adult person would bear the financial burden of one and three-tenths children. Those who rear fewer would be taxed, and those who rear more would receive a subsidy. Remember that all who rear children will have been determined to be competent.

To borrow Money on the credit of the (Country/Union), subject to the limitations provided herein;

To regulate Commerce with foreign Nations, and among the several (States/Provinces/Nations), to the extent that it is consistent with all of the limitations of this document; the government shall assess a tariff on all imported consumer products and services equally and without discrimination, which will be increased at least quarterly until the exports and imports are equal, after accommodating change in Net Currency Reserves. Such tariff shall be reduced to zero when imports of consumer goods and services are equal to or below the total exports.

This prevents the destabilizing disequilibrium caused by trade deficits that enhance consumption, which consumes capital assets and avoids the inevitable corruption that discretionary levies produce.

To establish a uniform Rule of Naturalization, and uniform Laws on the subject of Bankruptcies throughout the (Country/Union);

To establish a sound, reliable monetary and banking system and a quasi-independent central bank that can serve this role and that has a mandate limited to *"produce, provide, and preserve a sound, reliable, and stable money, currency, credit system, and appropriate accumulation of investment capital through regulating the required savings in the Personal Estate Fixed Investment Trust Accounts."* All who serve in this capacity can have no active current or future personal beneficial interest, excluding passive interests, in the banks or other institutions over which he/she has regulatory control. Controlled by this central bank, corporations and other commercial structures, including banks, other than investing in treasury securities, may borrow or lend only when secured by tangible assets and by not less than twenty percent equity. This provision would include deposits in banks. Broker/dealers and other financial institutions, which shall be separated from banks, may have the prerogative to post intangible collateral.

This will end the abusive and destabilizing practices of leveraging up the institutions, enabling the concentrations of power, and weakening their capital structures.

These limitations shall not apply to mutual benefit institutions/organizations. No bank or institution, the activity of which is considered to increase the supply or stock of money, shall engage in any other financial or commercial activity. Depository institutions shall carry insurance that sufficiently protects depositors. The Treasury Department shall issue currency and coin only to the extent requested by banks in order to meet the demands of the public. No real tangible assets may be mortgaged more than once; the loan must be amortized not to exceed the life of the asset, and be transferable with the lender's approval. Tangible improvements thereto are considered tangible assets and may be mortgaged in the same manner.

This is prescribed here to avoid the overpowering potential for the corruption of politicians through being granted such powers and to avoid the destabilizing impact of over-leveraging in an economy. This will assure greater stability to markets and strengthen capital structures.

To establish the Standard of Weights and Measures;

To provide for the effective deterrent and punishment of the counterfeiting of any securities or the currency or coin of the (Country/Union);

This would include "naked shorts" and other such counterfeiting schemes yet to be invented.

To promote the Progress of Science and useful Arts, by securing for a period not to exceed 30 years to Authors and Inventors the exclusive Right to their respective Writings and Discoveries;

To support scientific projects to an extent that does not exceed five percent of the Gross Domestic Product. Such expenditures will always be considered essential to the preservation of society and the people's sovereignty.

This is to ensure the advancement of society and to identify potential unknown threats.

To constitute Tribunals inferior to the Supreme Court;

To define and punish Piracies and Felonies committed on the high Seas, and Offenses against the Law of Nations;

To declare War, grant Letters of Marque and Reprisal, and make Rules concerning Captures on Land and Water;

To raise, support, provide, and maintain an Army, Navy, Air Force, and any other organization deemed necessary for defense, but no Appropriation of Money to that Use shall be for a longer Term than two Years;

To make Rules for the Government and Regulation of the land, air, naval, and any other defense Forces;

To provide for calling forth the Militia to execute the Laws of the (Country/Union), suppress Insurrections, and repel Invasions;

To provide for organizing, arming, and disciplining the Militia, and for governing such Part of them as may be employed in the service of the (Country/Union), reserving to the (States/Provinces/Nations) respectively, the Appointment of the Officers, and the Authority of training the Militia according to the discipline prescribed by Congress;

To engage any force necessary and without limitation in order to vanquish, subdue, or annihilate any threat to the functioning or survival of this Free-orderly society. No act in defense of this Free-orderly society shall be deemed too extreme or unjust.

To exercise exclusive Legislation in all Cases whatsoever, over and establish a District (not exceeding ten Miles square), as may, by Cession of particular States/Provinces/Nations and the Acceptance of Congress, become the Seat of this Government of the (Country/Union), consisting exclusively of Offices, Legislative buildings, monuments, memorials, and the residence of the President and no others, and to exercise like Authority over all Places purchased by the consent of the Legislature of the (State/Province/Nation) in which the Same shall be, for the erection of forts, magazines, other facilities for defense, arsenals, dockyards, and other needful buildings or facilities;

To make all Laws which shall be necessary and proper for carrying into Execution the foregoing Powers, and all other Powers vested by this Constitution in the Government of the (Country/Union), or in any Department or Officer thereof. In order to preserve the Sovereignty of the people, all laws and regulations promulgated hereunder not specifically required by this Constitution must be essential to the avoidance of chaos or gross and/or intolerable social or economic disorder and are necessary to preserve the remainder of the personal sovereignty of the people. All less significant matters of social order shall be reserved for the States and their inferior governments.

This is to avoid government and bureaucratic encroachment upon the sovereignty of the people and to set the bar very high when interpreting the word essential.

To dispose of and make all needful Rules and Regulations respecting the Territory or other Property belonging to the

(Country/Union); and nothing in this Constitution shall be so construed as to prejudice any Claims of the (Country/Union), or of any particular (State/Province/Nation).

To hold and preserve Lands, Monuments, and structures essential to the function of government;

To establish National Fixed Investment Trusts to hold assets and to distribute all income equally to the citizenry;

To establish a Land Trust to hold title to all productive lands, conduct public auctions of leases of such lands for an interval sufficient to amortize the required capital investment or fifty years, and to distribute the total net rents equally to the entire citizen population via National Fixed Investment Trusts. The NFIT shall distribute all income it receives equally among all the citizens of the district at regular intervals. All of the productive lands will be subject to the fifty-year land leases issued to the highest bidder at annual public auctions that each year will cover two percent of the total. The existing landowners' title will be converted to a fifty-year lease* and granted a First Lien equal to the depreciated replacement value of the improvements at the end of the lease. *(Initially, leases will be staggered through adding two-year increments by lot).

To establish Personal Estate Fixed Investment Trust accounts to assist in saving and forbearance of consumption.

Minimum personal savings must be established by the Central Bank, not to exceed fifteen percent. Personal Estate Fixed Investment Trust accounts may never be sold or liquidated. Personal Estate Fixed Investment Trust accounts shall never be pledged or attached and shall be impervious to any and all litigation. Personal Estate Fixed Investment Trust Accounts may only be transferred by Will, Probate, or gift to another Personal Estate Trust Account or a similar Trust for a not-for-profit organization.

An essential element of an economy to approach equilibrium is that the earnings of the productive assets flow to the consumers. This will also work to diminish the disparity of personal incomes and employ land at its most productive, highest use.

Section 2

The government shall be constrained as follows:

Except for laws governing Copyrights, Trademarks, Patents, and prohibiting fraud or deception, the Government or any inferior government shall make no laws creating impediments in market transactions, nor shall any such laws or regulations be adjudged valid or enforceable.

No Government Officers or employees may organize as a union or association.

No person or organization shall have the ability to coerce the Government.

The Privilege of the Writ of Habeas Corpus shall not be suspended, unless in Cases of Rebellion or Invasion, the public Safety may require it.

No Bill of Attainder or ex post facto Law shall be passed.

No Capitation or other direct Tax shall be laid, unless in proportion to or in relation to their total personal consumption.

No Money shall be drawn from the Treasury, but in Consequence of Appropriations made by Law; and a regular Statement and Account of the Receipts and Expenditures of all public Money shall be published at regular intervals.

This Government shall expend no monies in excess of revenues and currency expansion provided for in (Article IV Section 1) over a period of two years, except for the defense of the (Country/Union) or to cope with a natural disaster.

This is intended to avoid what experience has clearly taught: that government officials are universally inclined to spend more than they are willing to tax.

No person holding any Office of Trust shall, without the Consent of the Congress, accept any present, Emolument, Office, or Title from a foreign State, person, or other legal entity.

Government shall make no post-employment payments directly or indirectly to any officeholder or employee or guarantee employment to anyone. Persons who serve in the military, or families of such persons who were injured or killed in Line of Duty, and all Judges shall be excluded from this Section.

This is intended to avoid the temptation of government officials to provide unfunded compensation.

The Government may not engage in commercial activities or in economic activities that can otherwise be conducted by private ownership, except: it may own economically productive preserve land while leased to private interests for durations that allow for the amortization of improvements or a maximum of fifty years. Such leaseholders shall be entitled to compensation for the fair depreciated replacement value of the improvements at the time of the expiration of the lease.

Government-owned commercial enterprises shall be incorporated and the shares distributed on a per-capita basis to the benefit of all citizens via National Fixed Investment Trusts.

Government shall not tax or levy the Income, Production, or Property of any person, corporation, or other legal entity except as provided in Article V, Sec 1. A consumption tax shall not be construed as applicable in this Section. Taxes may be levied on real property only if the real property directly benefits from or causes the related governmental expenditure.

This is intended to prevent what has been the long and universal detrimental effect of income taxes. They have distorted and perverted economic and governmental decisions; they have corrupted governments and the governing process and undermined the integrity of taxpayers.

The Government shall not limit the manner of settlement of any contract between sovereign persons, be it in the domestic currency, foreign currency, any metal or commodity, or any agreed-upon compensation.

This deprives the Government of forcing a debased currency on its citizenry.

The Government shall not maintain, operate, or fund any prisons except for maintaining persons prior to adjudication. This provision shall become effective ***date*** ____. Addict Colonies shall not be defined as Prisons.

This is intended to force the criminal justice system to devise penalties that effectively serve to modify the behavior of the transgressors and to serve as an example that will deter others from engaging in the same or similar transgressions, as prisons have failed to do.

The Government shall not meddle in the affairs of other nations.

Provisions regarding Irresponsible Procreation are counter to the Sovereignty of the individual; both are essential to a Prosperous, Free-Orderly society, as both run counter to Human Nature with regard to procreation and the forbearance of consumption.

The goal here is for the income from the productive assets to flow to consumers—an essential function of an economy to approach a durable equilibrium.

Section 3

The government is mandated as follows:

All existing laws and each regulation thereunder as of this date shall, within a period of ten years, expire or be made to conform to this Constitution. Decennially after the passage of a law, it and all of the regulations thereunder promulgated will be reviewed by Congress and determined by a vote of the majority that the encroachment upon the individual's Sovereignty, as implemented, is essential and necessary for the preservation of their remaining Sovereignty in a Free-orderly society, or the law will expire.

This is to facilitate a smooth transition and to avoid the event of obsolete laws remaining in force.

The Government shall regulate non-governmental institutions to ensure their transparency and that they are free of fraud and deception. The Government shall mandate universal forms of disclosure for all products or services offered to consumers that will disclose all information for consumers to make rational decisions. This will include quantity, quality, identifiable risks, and the publishing of non-deceptive pricing. The Government shall regulate the commercial activity of enterprises operating in a natural monopoly. The Government shall not engage in regulating any commercial activity where consumer protection can be provided by a not-for-profit insurance or insuring institution deemed free of fraud and conflicts of interest.

The Government shall establish protections of the shareholders' interests in publicly owned corporations or companies having Boards of Directors collectively holding less than a majority interest; that require Boards of Directors have no fewer than seven members; that have Cumulative Voting; that prohibit Board members from being Officers or employees of the corporation or any related corporation; that assure preemptive rights; that enable

Shareholder Unions to individually seat Board members not to exceed twenty percent of any Board; that corporations have but one class of Capital – Common Stock of one class; that all such public corporations or businesses disburse no less than ninety percent of all net earnings as dividends; that incentive compensations be paid or disbursed as stock to Personal Estate Fixed Investment Trust Accounts. "Incentive shares" may never be exchanged and are not subject to the diversification requirements of the Personal Estate Fixed Investment Trust Accounts.

This constrains managers from serving their self-interests to the detriment of the public shareholders. This protects shareholder interests, makes corporate financial structures sound, assures the flow of earnings to the consumers—an essential economic function; and fosters longer-range corporate decision-making.

The Government shall: require the establishment of Personal Estate Fixed Investment Trust Accounts; require that all earned income be subject to saving and deposit to a Personal Estate Fixed Investment Trust Account of varying around ten percent in order to produce an investment aggregate yield of approximately five percent net of inflation established by the Central Bank; require that assets in the Personal Estate Fixed Investment Trust Accounts be selected by the owner; have initial investments not exceeding five percent of the total assets and that may be invested in Fixed Investment Trusts; and shall not be sold or liquidated but shall be transferable only to another Personal Estate Fixed Investment Trust Account or a Not-for-Profit organization. Assets of Personal Estate Fixed Investment Trust Accounts may not be used or pledged to secure any loans in any way. The Government shall charter Fixed Investment Trusts which may not sell or liquidate investments once acquired; that shall not acquire additional issues exceeding five percent of the total assets at the time of acquisition; that shall disburse all Net Income to the shareholders and never be reinvested.

Personal Estate Fixed Investment Trust Accounts shall be impervious to any legal action, attachment, or legislative action. The Government may establish Regional Fixed Investment Trusts for the residents of each region that are Personal Estate Fixed Investment Trust Account-eligible. The shares of corporations that produce seventy-five percent of their revenues in one or two regions may be offered to Personal Estate Fixed Investment Trust Accounts of persons residing in those regions.

All income taxes—both personal and corporate—and all estate or death taxes will be replaced by requiring all publicly held corporations to issue an amount equal to twenty-five percent of the shares outstanding after the conversion of debentures and any other unsecured debt to common stock to be distributed to National Fixed Investment Trusts (NFIT). All government-owned enterprises shall be capitalized exclusively with common stock, which shall be distributed to the National Fixed Investment Trusts. The NFIT will hold these shares and distribute all of the income equally to the legal adult population, and no assets in the NFIT shall ever be sold or liquidated. The National Fixed Investment Trusts have nothing to do with the Government.

This is intended to provide a mechanism through which earnings can be distributed across the entire population. This is to prevent all manner of imaginable corrupt activities.

While the mandatory asset accumulation—Personal Estate Fixed Investment Trust Accounts—provision and the...

(This sentence fragment appears to be incomplete and may need to be revised or completed.)

This is an essential structure for an economy to achieve near equilibrium. This will assure that each person shall eventually be financially secure, that there will be adequate capital available, and when combined with ending irresponsible procreation, each subsequent generation shall be economically better off than its predecessor, and provides an essential investment medium for the Personal Estate Fixed Investment Trust Accounts.

The Government shall prohibit any post-employment obligations on any government, company, corporation, or other employer except where provided in the Constitution.

The Government shall require that all newborns shall be protected through this provision that mandates the newborns shall only be permitted to a family or household in which two or more willing and competent adults, at least one of each gender, have entered into a mutual contract that provides that while this Constitution guarantees their total authority over the minor child, they collectively and severally accept total responsibility for the wellbeing of the child including the provision of food, clothing, shelter, healthcare, education, communication, and transportation for him/her until he/she attains adulthood as defined by the Government and to a standard determined by the Government. To bear or beget a newborn not permitted to a qualified family or household shall be subjected to confiscation of the newborn and sterilization of the parents.

A Free-orderly Society cannot coexist with irresponsible procreation. Every newborn deserves and requires the care of competent adults. The number of which, over two, is of no consequence as long as the willing and competent adults have entered into a mutual contract to provide for the newborn to a specified level until adulthood.

The Government shall require that each immigrating person shall be able to provide an income for, or be provided for, to an extent at least equal to the level of income established by the Congress that is adequate to provide food, clothing, shelter, healthcare, education, communication, and transportation for the individual and all dependents for a foreseeable period specified by Congress. Work visas not exceeding one year may be issued and are exempt from the previous conditions.

No Tax or Duty shall be laid on articles exported from any (State/Province). No Preference shall be given by any Regulation of Commerce or Revenue to the Ports of one (State/Province) over those of another; nor shall Vessels or Land transport bound to or from one (State/Province) be obliged to enter, clear, or pay Duties in another.

Full Faith and Credit shall be given in each (State/Province/Nation) to the public Acts, Records, and Judicial Proceedings of every other (State/Province/Nation). And the Congress may, by general Laws, prescribe the Manner in which such Acts, Records, and Proceedings shall be proved, and the Effect thereof.

New (States/Provinces/Nations) may be admitted into this (Country/Union) with the approval of two-thirds of the States; but no new (State/Province/Nation) shall be formed or erected within the Jurisdiction of any other (State/Province/Nation); nor any (State/Province/Nation) be formed by the Junction of two or more (States/Provinces/Nations), or Parts of (States/Provinces/Nation), without the Consent of the Legislatures of the (States/Provinces/Nations) concerned, as well as of the Government.

The (Country/Union) shall: guarantee to every (State/Province/Nation) in this (Country/Union) a Republican Form of Government; encourage (State/Province/Nation) to style a Constitution after this document; and shall protect each of them against Invasion, and, on Application of the Legislature, or of the Executive (when the Legislature cannot be convened), against domestic Violence.

The powers not delegated to the (Country/Union) by this Constitution, nor prohibited by it to the (States/Provinces/Nations), are reserved respectively to the sovereign people or to the (States/Provinces/Nations).

All persons born of Citizens or naturalized, and subject to the jurisdiction thereof, are citizens of the (Country/Union) and of the (State/Province/Nation) wherein they reside. No (State/Province/Nation) shall make or enforce any law which shall abridge the privileges or immunities of citizens of the (Country/Union); nor shall any (State/Province/Nation) deprive any person of life or sovereignty over his/her person, property, or minor children, without due process of law; nor deny to any person within its jurisdiction the equal protection of the laws.

The Government shall prohibit any combinations, associations, or non-natural monopolies affecting any basic or essential economic activities and shall not create impediments in the markets for essential economic activities.

Excluded from this provision, Labor Unions may organize to the extent of one employer in one market area.

Business organizations have two classes of employees – management employees and operational employees – whose interests are in conflict. Therefore, the ability of each to negotiate their respective interests is herein guaranteed, but only to the extent of their respective and common interests – one employer in one market. If "the ship sinks, they all go down together." Also, the Union cannot sacrifice some members for the benefit of others to influence their total market.

Nothing in this Constitution shall apply to the manner in which (States/Provinces/Nations) engage those who the (State/Provincial/National) judicial system has determined to be habitually harmful to the person, property, or minor children of others.

This is intended to allow creativity in the manner that the incorrigible are dealt with by the (States/Provinces/Nations).

The Congress shall make laws that promote, protect, and preserve each individual's sovereignty from intentional encroachment only by other individuals. The Congress shall make laws that enable equitable settlement in cases where unintentional encroachment upon the sovereignty of, or harm to, the person, property, or minor children of others occurs.

This restricts the reach of the Government into the lives of the Sovereign citizens.

The Government shall make laws that protect the practice of all religions that can be practiced in a manner that respects the sovereignty of each person over his person, property, and minor children, and all other religions that can be practiced in the same manner.

The Government shall provide that no commercial transactions shall be conducted with a minor except through a minor acting as an agent of a parent or guardian, and shall comply with the expressed authority granted to the minor by the parents or guardians.

This assures the sovereignty of the parents over their minor children. The parents can provide children with electronic debit cards that may have restrictions on the amount of a transaction and/or provide for specific prohibitions of transactions for such things as alcohol, sugar, tobacco, pornography, restricted-rated entertainment, and such.

For a period not to exceed fifty years after the signing of this Constitution, the Government may provide an economic "safety net" for citizens that would provide for basic sustenance rations of food, clothing, shelter, transportation, medical care, communication, and education through vouchers for each paid from the General Fund.

Being distributed in the form of vouchers only will preempt and replace a plethora of programs that serve special interests, pervert the markets, and corrupt government. Establishment of charitable eleemosynary organizations to provide such services will enable them to continue providing such services after the government assistance ends.

The Government shall assume financial responsibility for all education as follows: Commencing two years after the signing of this Constitution, ten percent of the mean average per-student costs to State and local governments and charged by private providers during the previous five years in the market area. This amount will increase by two percent each year for twenty years. This shall be disbursed without discrimination in the form of vouchers to the adults responsible for the education of children or themselves. These vouchers may be made contingent on "Good Conduct." Good Conduct is that behavior that does not harm or coerce others.

This will gradually relieve the (States'/Provinces'/Nations') burden, shifting it to the Government. After the prescribed period, the individual and society will bear equally the burden for the respective benefit to each. (See Article VII Section 6.) The vouchers are an essential method of payment, as any other would foster corruption. Great adjustments and changes in education will evolve over the period. Paid for through VAT (consumer tax), the greater money benefits to the richer regions will be offset by the greater tax paid in those regions. Society and the educated person hold education as a capital asset.

For a period of twenty years after the adoption of this Constitution, there will be imposed several provisions by addendum, known as THE RECOVERY AND IMPLEMENTATION PERIOD, that will supersede any provisions of this Constitution.

Article VI. Limitations and Responsibilities of (State/Province/Nation)

Section 1

No (State/Province/Nation) shall enter into any Treaty, Alliance, or Confederation; grant Letters of Marque and Reprisal; coin Money; emit Bills of Credit; make anything but gold and silver Coin a Tender in Payment of Debts; pass any Bill of Attainder, ex post facto Law, or Law impairing the Obligation of Contracts.

Section 2

No (State/Province/Nation) shall borrow or sell General Obligation securities having a maturity of greater than thirteen months or other securities that are not secured by tangible assets equal to one hundred and ten percent of the borrowing.

Section 3

No (State/Province/Nation) shall, without the Consent of this Government, lay any Imposts or Duties on Imports or Exports, except what may be absolutely necessary for executing its inspection Laws; and the net produce of all Duties and Imposts laid by any (State/Province/Nation) on Imports or Exports shall be for the use of the Treasury of the (Country/Union); and all such laws shall be subject to the revision and control of the Congress.

Section 4No (State/Province/Nation) shall, without the Consent of this Government, lay any Duty of Tonnage, keep Troops, or Ships of War in time of Peace, enter into any Agreement or Compact with another (State/Province/Nation), or with a foreign Power, or engage in War, unless actually invaded, or in such imminent Danger as will not admit of delay.

Section 5

No (State/Province/Nation) shall, without the Consent of this Government, undertake jurisdiction of any activities that are provided for here as being in the domain of this Government.

Section 6

No (State/Province/Nation), or its inferior governmental bodies, shall make laws that do not conform to the intent, protections, and provisions of this Constitution.

Section 7

Twenty-two years after the date of this Constitution, no (State/Province/Nation), or sub-governing bodies, shall own, operate, or manage schools or universities, or engage in any commercial activities.

Section 8

No (State/Province/Nation), or its inferior governmental bodies, shall impose income taxes, estate taxes, allow unions of state employees, or provide post-employment compensation to officials or employees.

Section 9

Any (State/Province/Nation) may enact legislation that is deemed essential to the preservation of the remainder of individual sovereignty when dealing with matters and conditions peculiar to its jurisdiction.

Article VII. Expressed Protections of the Sovereign People

Section 1

The Citizens of each (State/Province/Nation) shall be entitled to all privileges and immunities of citizens in the several (States/Provinces/Nations). A person charged in any (State/Province/Nation) with treason, felony, or other crime, who shall flee from justice and be found in another (State/Province/Nation), shall, on demand of the executive authority of the (State/Province/Nation) from which he/she fled, be delivered up to be removed to the (State/Province/Nation) having jurisdiction of the crime.

Section 2

All adult persons shall be considered sovereign – possessing total authority and total responsibility – over and for their person, property, and minor children, and no law shall abridge this sovereignty except as provided for by this Constitution and unless the person is adjudged incompetent by a court of law.

Section 3

Excluding the philosophy established here of a Free-orderly society, the Congress shall make no law respecting an establishment of religion or inhibiting the practice or observance of any religion that can be practiced and observed in a manner that respects the person, property, and minor children of all people; their sovereignty over the same; and the practice of all other religions that are practiced and observed in the same manner.

Section 4

Except where imminent physical harm may result, the Congress shall make no law prohibiting or abridging freedom of speech, or of the press or communication media, or enabling personal communication media to mute legitimate communication, or the right of the people peaceably to assemble within an enclosure free of coercing others and to petition this Government for a redress of grievances. Such a petition may only be made by any person on their own behalf or on behalf of any legal person, conducted by a single person. No such agent representative shall have served the Government in the preceding nine years.

Three issues are addressed here: The "yelling fire" issue, or any speech that would cause physical harm to others, is a necessary restraint on speech. Assemblies of persons for the purpose of producing coercion are incompatible with a Free-orderly society. And the lobbying of members of Congress by armies of lobbyists is incompatible with an unencumbered government dedicated to producing, providing, preserving, and protecting a Free-orderly society.

Section 5

The right of the Sovereign people to bear arms for the purpose of self-defense, protection of their sovereignty, and hunting shall not be infringed.

Section 6

The personal sovereignty of the people shall grant their right to secure their persons, houses, papers, records, electronic memory, and effects against searches and seizures and shall not be violated but upon consent, specific or by deed. Violation of this provision is tantamount to self-incrimination. No personal act in defense of this right may be challenged.

Searching one's personal papers and electronic memory is as searching one's personal memory — via torture.

Section 7

No person shall be held to answer for a capital, or otherwise infamous, crime unless on a presentment or indictment of a Grand Jury, except in cases arising in the land or naval forces, or in the militia, when in actual service in time of war or public danger; nor shall any person be subject for the same offense to be twice put in jeopardy of life or limb; nor shall be compelled in any criminal case to be a witness against himself; nor be deprived of life, liberty, or property without due process of law; nor shall private property be taken for public use without just compensation; nor shall be prosecuted for falsehoods except those issued under oath to be truthful, or falsehoods known as such published with consent, or the presentment of false identification.

Section 8

In all criminal prosecutions, the accused shall enjoy the right to a speedy and public trial by an impartial jury of the (State/Province/Nation) and district wherein the crime shall have been committed, which district shall have been previously ascertained by law; and to be informed of the nature and cause of the accusation; to be confronted with the witnesses against him; to have compulsory process for obtaining witnesses in his favor; and to have the assistance of counsel for his defense. All evidence available to the prosecution shall be made available to the defense. Failure to do so shall be a crime, and the Government shall provide for the effective deterrent and punishment of such. All accused shall be considered innocent until adjudicated guilty. Each such public trial shall be concluded within one year at any level of judicial jurisdiction. Failure shall be cause for effective punishment of any and all judges and officers of the court of jurisdiction.

Punishments delayed fail their purpose. Withholding exculpatory evidence shall be the equivalent of "bearing false witness."

Section 9

To provide a Free-orderly Society, all punishments must have a high rate of effectiveness in modifying behavior and/or deterring unacceptable behavior at minimal public expense and do so while excluding imprisonment paid for with public funds as an option. Punishments must be determined by a panel of psychiatrists and psychologists.

Punishments for the incorrigibles must remove them from the (Country/Union) by relocation outside of the (Country/Union), or by death. Castration shall be an acceptable punishment for the purpose of modifying the behavior of convicted sex criminals or for crimes of habitual aggressive behavior. Punishments intended to deter potential criminals must be conducted publicly. Excessive bail shall not be required.

The intent of Section 9 is to make clear certain matters of punishments so that Congress or the Courts will not stray, making punishments ineffectual. Ineffectual punishments harm all sovereign individuals. Imprisonment constitutes the coercion of sovereign individuals to provide monies for the care and feeding of the least worthy in society. Imprisonment has proved ineffectual in modifying behavior and is a punishing cost to society, both in terms of maintaining those incarcerated and the cost of the loss of their production if behavior modification were otherwise accomplished. The alternative to imprisonment has stultified creativity regarding methods of behavior modification.

Section 10

In suits among and between adult persons, where the value in controversy shall exceed the price of one hundred ounces of gold, the right of trial by jury shall be preserved, and no fact tried by a jury shall be otherwise reexamined in any court of the (Country/Union) than according to the rules of the law. All other such suits shall be settled in a fair, equitable, and efficient manner as provided by law. All persons found to be innocent of any complaint, criminal or civil, shall be reimbursed for all the costs of his/her legal defense.

The two issues dealt with here are: (a) that specifying a threshold in terms of a currency becomes inappropriate over time, and (b) innocent persons must be unharmed as a result of any litigation.

Section 11

The Government shall provide every person with a simple identification card containing a photograph, definitive biometric identification, or an electronic device containing a unique assigned number and a resident address.

Section 12

No Minor may be brought before any Court except as a witness and with the approval of the responsible adult or adults. Those adults responsible for the minors must answer to the Court for the acts of the minors for whom they are responsible.

Section 13

All sovereign persons may petition the Courts challenging any law or regulation as "essential to the preservation of the remainder of their sovereignty."

Section 14

The enumeration in this Constitution of certain rights shall not be construed to deny or disparage others retained by the sovereign people.

Section 15

No sovereign person may be convicted of a crime of which he/she had no knowledge.

Punishments that effectively modify behavior are frequently too harsh to be applied to a person whose behavior is not subject to modification. Only those who knowingly commit a crime should be subjected to modification.

Article VIII. Responsibilities of the People

Section 1

All sovereign persons who remain in the jurisdiction of this Constitution by this act endorse this Constitution and thereby agree to conduct themselves in a manner that honors the sovereignty of all others over their Person, Property, and Minor Children and the practice of all religions that can be conducted in a manner that respects the Person, Property, Minor Children, and the practice of all other religions.Here lies the essence of the morality of a Free-orderly Society.

Section 2

A Free-orderly society requires all Sovereign people to assume all responsibility for themselves, their property, and their minor children. To ensure that no person shall be denied the opportunity to fulfill these essential responsibilities, the government shall ensure the availability of insurance and/or financial organizations or private or not-for-profit insurers to provide for sustenance in old age/disability, basic medical care, protection from income disruption, and accidental injury to others.

Section 3

Except for the provisions of Section 2 of this Article, persons who are habitually and incorrigibly irresponsible, for whatever reason as properly determined by a Court of Law, in matters of their Person, Property, or Minor Children may become a ward of the (State/Province/Nation) and be dealt with in any manner that the (State/Province/Nation) government may determine.

Such persons present a problem to the local community, which may deal with the problems free of (Country/Union) interference.

Section 4

To assure the continuation of this society, Governments may mandate that each adult accept financial responsibility for the rearing of children, either through the rearing of children or through taxation, sufficient to create a number of permitted children to provide for society's continuance. In order to maintain an adequate population, each adult may be required to be financially responsible for the rearing of the whole and fractional number of children. The amount shall be based on and assessed from the average cost in each (State/Province/Nation), and be paid either in-kind or as a tax, made age-appropriate for the payer and the cost of child-rearing at those ages in that market area. Proceeds of the tax shall be distributed on the same basis to those families that exceed that requirement.

Section 5

The portion of the cost of education of the children borne by the parents or guardians shall commence at one percent and increase annually by one percent for forty-nine years.

A portion of education is a "capital asset" of the holder, while a portion is an asset to society; therefore, it is provided here that each shall share the burden. As the government's and the parents' share of the financial burden of education increases, the burden on the state and local governments will disappear. As all education decisions are made by the parents, market forces will prevail.

Article IX. Continuity of Government

All Debts contracted and Engagements entered into before the Adoption of this Constitution by any previous (Country/Union) shall be subject to validation or repudiation by the Congress of the (Country/Union) under this Constitution.

This Constitution, and the Laws of the (Country/Union) which shall be made in Pursuance thereof; and all Treaties made that conform to this Constitution, or which shall be made under the Authority of the (Country/Union), shall be the supreme Law of the Land; and the Judges in every (State/Province/Nation) shall be bound thereby, anything in the Constitution or Laws of any (State/Province/Nation) to the contrary notwithstanding.

All existing Laws and Treaties, as of the date of this Constitution, shall prevail for one year after a determination of nonconformity to this Constitution by the Supreme Court, enabling Congress to replace or modify the non-conforming Law.

The Senators and Representatives before mentioned, and the Members of the several (State/Provincial/National) Legislatures, and all executive and judicial Officers, both of the (Country/Union) and of the several (States/Provinces/Nations), shall be bound by Oath or Affirmation to support this Constitution; but no religious Test shall ever be required as a Qualification to any Office or public Trust under the (Country/Union).

Article X. Amendment

Amendments to this Constitution and the subsequent addition of (States/Provinces/Nations) may be initiated by Congress or any (State/Province/Nation) and shall require the consent of seventy-five percent of the (State/Provincial/National) Legislatures.

Be warned: any attempts to amend this Constitution, prior to or subsequent to its adoption, will most likely be motivated by efforts to serve some special interests of some at the expense of others, until proven to the contrary.

Article XI. Pledge of Allegiance

All persons who are residents of this (Country/Union), after a period of one year, shall be deemed to have expressly subscribed to, endorsed, and pledged supreme allegiance to this Constitution by their act of remaining in residence and thereby expressly or implicitly made an oath to respect this Constitution as follows:

"I pledge my solemn oath to assume total responsibility for my person, property, and minor children, and to defend, protect, and preserve the (Country/Union) and my supreme allegiance to its Constitution, which I shall at all times honor and by which I shall abide."

Article XII. Ratification and Endorsement

The ratification of the Conventions of (number) (States/Provinces/Nations) shall be sufficient for the establishment of this Constitution between the (States/Provinces/Nations) so ratifying the same.

Attest ____________________ Secretary

The society that adopts a constitution that faithfully honors this outlined Constitution will enjoy durable prosperity, freedom, and order.

OTHER WRITINGS OF THE INSTITUTE OF PERCEPTIONISM®

Proposed Amendments To US Constitution

The Institute of PERCEPTIONISM®

Offers this outline of the U.S. Constitution Amendments to achieve a

Free-Orderly, Durably Prosperous Society

1a. Sovereignty of Each Individual Adult Citizen (The Freedom Amendment)

Preamble:

Human beings are created to develop into freestanding, self-contained, independent, mobile entities. To that extent, with few exceptions, we are all equal. To be endowed with Rights to Life, Liberty, and the pursuit of Happiness implicitly requires that each adult individual is autonomously sovereign over their person, property, and minor children. To that extent, we are free. To whatever extent we are not sovereign over our person, property, and minor children, someone else is—and we are their slaves.

Because of the vast range of natural human abilities in so many different fields and activities, and the capriciousness of good or misfortune, the differing outcomes on every level for each individual will be great. It must be recognized that, excluding reproductive attributes, virtually all traits commonly attributed to women appear, in varying degrees, in all men, and that traits commonly attributed to men appear in all women to a varying degree. Even if two people on Earth were to have the same physical being, neither would have the same brain circuitry when combined with environmental conditioning. All seven billion plus of us are truly different and unique individuals.

To propound anything to the contrary is an extremely harmful prevarication. *

That which is in italic is an explanation only.

It is and has been this capricious engagement among the human population that has created the phenomenal advancement of the species. The allowing and rewarding of the free/orderly competition, breeding, nurturing, educating, and endeavor of individuals has and will produce great human achievement. This foundation of societal design assures infinite improvement of the species.

To this end, all laws must honor and be compatible with these self-evident truths about the human condition.

Therefore:

To achieve a Free-Orderly society, this Government shall provide, preserve, and protect the sovereignty of each adult citizen over their person, property, and minor children—that is, absolute life-and-death authority over, and total responsibility for, one's person, property, and minor children. Any compromise of this personal sovereignty must satisfy the test that the compromise is essential for the survival of the remainder of individuals' sovereignty.

To this end, the Government assures that each and every person shall conduct themselves in a manner that honors the sovereignty of all others and does no harm to any others' person, property, or minor children without reparation or remunerated damages. Harmful social intercourse shall be between or among the affected parties, guided by the laws of compensation. Only repeated harmful behavior, violent behavior, or violation of law or human activity that is incompatible with a free/orderly society shall fall within government purview.

Speech, to be considered harmful, must cause specific, measurable, tangible damage.

An assembly for the purpose of advancing a point of view must be contained within an enclosure. An assembly that exposes unwilling other sovereign persons to coercion is an unacceptable condition in a free/orderly society. Coercion is the antithesis of freedom.

A "People's Court" Judge shall hear all disputes between sovereign parties, make a judgment, and remand discontented parties to a higher court, to arbitration, to special instruction and counseling, or to the State's Attorney. The People's Court Judge shall hear all misdemeanors including civil disobedience. After the third offense, the case will be remanded to the State's Attorney.

Persons who conduct themselves in a manner that harms others—for first or second-time offenses—will undergo instruction in behavior modification paid for or reimbursed by the offender, along with any damages, with payment styled to be within their ability to pay. The government shall assure that any and all persons who demonstrate, by a third intentional harming of the person, property, or minor children of others or other adjudicated repeat offenses, shall be deemed incorrigible and unable to behave in a manner consistent with these principles, and are removed from the society without burden to those persons who do behave within these parameters, by a sentence to death or by obtaining residence in another jurisdiction. Premeditated murder shall be punished by death, as will third-time offenders who do not find residence in some other jurisdiction.

Any nation that protests the death penalty needs only to supply a visa. Prisons, therefore, are not an option. "Prison time" pays no debt to society. It only financially burdens the orderly persons of society. Offenders must pay those who are damaged, even if only to an amount that, after payment, the offender can survive.

All law enforcement officers will be provided total indemnification. Law enforcement officers are agents of their employer. Therefore, the employer is totally responsible for hiring and training the officers. Any failure in their performance while on duty rests solely with the employer.

For this society to be free and orderly, whenever or wherever life and freedom are in conflict, life shall yield. Wherever and whenever life is more important than freedom, freedom will eventually vanish—as the preservation of life has no limit.

Lethal, addictive drugs are banned. Purveyors shall be sentenced to death. The third conviction for intentional possession will cause the person to be sentenced to death or successfully depart from the larger jurisdiction. Drug addiction threatens the integrity of a free/orderly society.

The Government shall not have access to one's electronic memory devices without permission from its owner.

One's electronic memory and organic memory must be treated equally to enable a prohibition of self-incrimination.

The Government shall provide and enable institutions of commerce and charity that facilitate free, honest, enforceable contractual interaction and a free, honest marketplace among the sovereign people as they satisfy their needs and desires in their pursuit of happiness. All purveyors of goods or services shall publish, prior to any transaction, a full, honest, forthright disclosure including price, fees, and/or charges. The Government shall facilitate independent, honest rating services and methods to aid consumers in evaluating where they spend their money, energy, or time. The Government shall not regulate or control purveyors in the marketplace beyond these mandates. Natural monopolies shall fall within the purview of the Government.

Neither the Federal nor the State Governments shall engage in any enterprise that can be performed by private enterprises in a competitive marketplace.

Any power in the hands of politicians, bureaucrats, or dictators will be corrupt and perverted.

Coalitions, associations, combinations, or unions that function to pervert prices in the marketplace are banned. Excluded are institutions that make market participants better informed.

Only the ability to perform the task under consideration may be considered when selecting a person to fulfill a task or position. Factors such as race, national origin, family, or gender of a sovereign person may never be considered.

The Government, within that which is in its purview, shall protect this society from all that threatens its continuance as a free-orderly society.

The ratification of this Amendment shall constitute an amendment of the State's Charter and Constitution.

Each of the succeeding Amendments shall specify a date not to exceed ten years by which all laws will conform to the provisions of these Amendments. Whenever and wherever these Amendments are in conflict with the existing Constitution and any interpretations thereof, these Amendments shall prevail.

By this Amendment, all power becomes vested in the governed citizenry. This is the lowest common denominator of freedom, and it is enjoyed equally among all citizens. Any compromise of that sovereignty is a move in the direction of sublimation to tyranny, serfdom, and/or slavery, for which there is no limit.

Any effort to provide for care and compassion in this social structure must, therefore, be pursued through pure charity.

Democracies are extremely tyrannical and implicitly corrupt, bearing little resemblance to a truly free/orderly society.

2a. Reestablishment of the Sanctity of Private Property

Preamble:

The sanctity of private property and ownership of one's production is fundamental to individual freedom and assures the true freedom of each individual citizen.

Therefore:

This Amendment repeals and replaces the 16th Amendment.

All income taxes—both personal and corporate—along with all estate or death taxes and all discriminating excise taxes are banned and replaced by value-added taxes or sales taxes uniformly applied to all consumer goods and services, and/or by tariffs, and by fees for government services and the use of government properties.

The cost of Government is a part of every resident's consumption and is financed by value-added tax or sales taxes uniformly applied to all goods and services, ensuring that all pay. Those who consume the most are presumed to receive the greatest benefit and, in this manner, pay a commensurate tax and thereby share a relatively equal burden.

There is no discernible difference between the confiscation of one's production or one's property by the Government, by the Noble (serfdom), or by the plantation owner (slavery).

3a. Reestablishment of the Integrity of the Republic

Preamble:

States and local governments deal with differing problems. States cannot be burdened by problems unique to other States. The natural desirability, climate, and multiplicity of natural resources and other features make one State more desirable to different people than others at different points in time. Any attempt to make them equal would make the citizens of other States slaves to the benefiter

State. Each State acts to provide, protect, and preserve the sovereignty of its citizens as it deals with its unique conditions. The Senators are not to represent households but rather all of the citizens of their unique States, as determined by their Legislatures. By State governments dealing with the problems unique to their region, corruption of concentrated power is reduced while needs are met.

Therefore:

This repeals and replaces the 17th Amendment.

The States shall adopt the provisions of the Freedom Amendment and cause their constitutions and laws to be compatible with the remainder of these amendments within the same time period established for implementation of these amendments.

4a. Provides terms, conditions, and requirements for all office holders and other government employees.

Preamble:

This avoids the inefficiencies of unnecessary elections; the professional politician, and assures that all office holders have had a modicum of life experience and, hopefully, honorable intentions. The role of the professional bureaucrat will be profoundly diminished in a society where responsibility and authority are vested in the sovereign individual citizens—not in government. As a result, the officeholder will be less dependent upon the bureaucrats, so that the need for governing experience among the officeholders is minimal. The role of the elected officials becomes that of custodians and trustees who must constantly guard against the propensity to increase their power. It is long understood that power corrupts. Human beings almost universally find, upon obtaining a position of power at any level and in every place, their thoughts drifting to how they may apply this power for their personal benefit. All institutions must be styled in a way to minimize this very human, corruptive trait. The most successful way is to disperse and limit the power of any individual.

Therefore:

This repeals and replaces the 22nd Amendment.

The President shall serve one six-year term, and each new officeholder shall serve no more than one six-year term in a lifetime in any one classification of office. All offices shall have terms of six years. The Congress and the Senate shall be divided into three two-year classes.

No federal officeholder shall be less than forty-five years of age or greater than seventy years of age and shall receive no compensation other than reimbursement for authorized expenses. They will serve for the honor and the personal fulfillment of high public service.

Political parties or any similar voter-organizing organization shall be banned. This shall include public employee unions.

Political parties are coalitions of conspirators who conspire to use the coercive power of government to make laws that benefit themselves or their causes at the expense of other citizens—the antithesis of individual freedom. With most power vested in the individual, that power which falls to government will be minimal. This amendment enables a system of competent representation without political parties.

Qualification for running for an office of government will only require endorsement by no fewer than five not-for-profit public service organizations that serve the entire district, state, province, or nation served by that elective office. This will provide vetting of those seeking office.

If the government fails to achieve balanced budgets across three consecutive fiscal years, general elections are mandated for all offices within three months.

Any enterprise subject to government regulation or that supplies goods or services to the government, which hires a person previously employed by the department of government so involved or provides any personal emolument, will be fined an amount equal to 500 times the person's total compensation.

5a. Government Serves and Honors the Sovereign Family

Preamble:

The evolution of humanity has created the family as the foundational societal building block, as the family provides the nurturing, psychological maturation, socialization, education, economic well-being, healthcare, and compassion of its members. The government has not been able to successfully perform these functions. The government cannot provide care and compassion for some without violating the sovereignty of others. This avoids the conscription or enslavement of one group or class of citizens to serve the well-being of another.

Therefore:

This amendment repeals and replaces the 19th Amendment.

Each household shall have one vote, with no regard to gender. This will assure that the interests of the sovereign families are reflected in that which is in the purview of government.

Each family is responsible for matters of "care and compassion." Government shall not be involved with matters of "care and compassion," with the only exceptions being for persons injured in uninsurable natural disasters or in public service.

Henceforth, matters of care and compassion will be relegated to the PEFITA-endowed family, the multitude of not-for-profit service organizations, disease-specific foundations, and insurance companies.

Because of the disproportionate requirement for expensive technological healthcare services, the government shall subsidize a market-disciplined healthcare option for all people in residence through a Value Added Tax-funded Trust. (See Perceptionism Healthcare below.)

As a result of healthcare evolving into a major portion of human consumption—having no apparent technological upper limit—a market-disciplined limit and a universal subsidy are required. This market-disciplined subsidy must be self-governing, free of other human influence.

Care and compassion have no limit. Governments that assume responsibility for care and compassion have accepted an impossible task and will fail in the eyes of any population. To try is to enslave one portion into the service of another.

6a. Responsible Procreation

Preamble:

Irresponsible Procreation is totally incompatible with a *Free-orderly* society. This compromising of individual sovereignty is essential to the existence and a positive evolution of humanity in a *Free-orderly* society. If humanity has any purpose or value, *Responsible Procreation* is the highest achievement of most human beings. This is to ensure that every newborn starts life with at least two competent, contractually committed adults. To procreate without the willingness and ability to nurture and rear a child is to assume an authority without the commensurate responsibility, which is, for the newborn, the rankest form of child abuse. Inadequate parenting produces adults having debilitating psychological problems.

The evolution of the human species in the nuclear family has imprinted this structure on the human brain and psychological behavior. The dishonoring of this fact manifests itself in gross, erratic, antisocial behavior.

Therefore:

The State shall issue a *Permit* for each baby intended to ensure that every baby will go to its new home with at least two competent adults who have entered into a *Family Contract* to mutually and severally nurture, educate, and provide a loving, caring, developmental environment until adulthood. This requires that a mother must have the singular occupation of mothering each child for its first three years of life. Each partner of this *Family Contract* will have attained some minimum level of competence in order to qualify for a *Permitted Child.*

Qualification will require certain educational achievements for the skills of a functional family, the bearing and nurturing of children, and the occupational skills necessary to support the family. At least one member shall be engaged in a productive activity that enhances the human condition, demonstrating a justification for their replacement.

Entered into as sovereign individuals, the *Family Contract,* enforced by the States, shall stipulate all of the terms and conditions of the relationship, including a procedure for dispute settlement – effectively, the *Family Constitution.* The *Family Contract* shall supersede all secular "marriage law."

If reverence for human life is to be preserved, conception and birth must be looked upon as life's fulfillment for the years of self-denial and forbearance necessary for the joy of achievement by competently nurturing, rearing, and educating a child for a productive role in society.

In a society where all receive sufficient income for survival, only those couples having a genetic propensity to be engaged in a productive occupation shall have the permission to procreate. Persons with no contributory role need not be replaced.

The significance of the evolutionary role of each gender in the psychological development of a child will likely be discernible at a time after different family structures have been experienced in several states. Combined with PEFITAs, *Responsible Procreation* will rapidly diminish social welfare programs.

The Government shall establish protections of the shareholders' interests in publicly owned corporations or commercial associations having Boards of Directors collectively holding less than seventy percent equity interest. These protections shall require that Boards of Directors have no fewer than seven members; that there shall be Cumulative Voting; that Board members shall be prohibited from being Officers or employees of the corporation or association or any related association or corporation; that shareholders shall have preemptive rights; and that *Shareholder Unions* may individually seat Board members not to exceed twenty percent of any Board. Passive investors shall have no votes directly or indirectly for Directors.

The government will maintain a categorized registry of applicants desiring to serve as corporate Board of Directors members, who list qualifications and required compensation. Prospective Board members may be nominated one each by management, the employee union, the Board of Directors, or the number of shareholders, or their proxy, required to seat a board member with cumulative voting. There will be no fewer than two nor more than four nominated for each seat to be elected by the shareholders. Directors shall serve a term of six years, barring malfeasance or misfeasance, in three two-year classes. Eligibility to run for reelection requires that the company have had an average annual increase in earnings of at least two percent, adjusted for inflation, during its six-year term. The directors' total compensation must consist of at least 25 percent in company shares deposited to their PEFITA.

These mandates will establish and reestablish provisions that avoid a plethora of corporate management abuses of their powers that serve the managers' personal interests at the expense of the shareholders.

The custodians of the assets of the passive investors shall have no voting rights, entitlement, or responsibility to change their passive condition.

All corporations or commercial associations shall have only one class of capital – *Common Stock* or equity interests of one class. All corporations or businesses that have had public shareholders shall disburse all net income, not invested in the company's productive capital assets, as cash dividends and never engage in buying back the corporation's stock. Additional capital requirements will be supplied by the issuance of additional common shares, first offered to existing shareholders.

This will avoid destabilizing leverage and enable broader, less concentrated ownership of corporate shares, and the flow of earnings of the capital assets to the consumers, as the major supplier of capital will be PEFITAs. This is a major step in the direction of reducing the great disparity of wealth and income.

All institutional lending, other than banks, will be fully amortized and secured by 120% tangible assets or by investment-grade guaranteed deposits or intangible assets having a maturity of less than one year. Upon satisfaction, a loan is extinguished, and with the exception of a succeeding homeowner, the asset may never be re-collateralized again. Improvements to existing assets will be considered new assets.

Bank lending may accept financial assets as collateral at no less than 200% of the loan amount for market-making institutions. Banks shall maintain twenty percent capital reserves. This strengthens the banking system and minimizes the leverage and speculative activity that is responsible for much of the cycle of economic instability.

Commercial banks shall not engage in underwriting, making a market, or trading in common stock, nor may any officer of a commercial bank serve as an officer or director of an entity that does. This reestablishes the provisions of the *Glass-Steagall Act* that were successful for sixty-five years.

Incentive compensation must be generous enough to obtain and motivate the best talent to their highest performance. All Marxist systems and bureaucracies have demonstrated beyond any doubt that proper incentives for effort and achievement are critical for any successful system.

Incentive compensations paid or disbursed as stock shall be to *Personal Estate Trust Accounts* as "Incentive shares" in a PEFITA, which may never be exchanged and thereby are exempt from diversification requirements of the PEFITA. This stock incentive program encourages, if not assures, longer-term corporate planning.

Institutions of age-specific retirement, either public or private, shall not be allowed. Retirement shall be a decision made by each individual with their employer. Institutionalized pensions are commonly contractual arrangements made by and for one generation to be paid by the next. This is untenable because it burdens the subsequent generation into the service of its predecessor, over which the latter had no input or control.

The *Personal Estate Trust Accounts* will eventually replace most retirement needs; however, it is not the entire capital base. There will be normal markets for vast assets not held in the *Personal Estate Trust Accounts* to be acquired for retirement security.

Labor unions shall be limited to one employer in one "market area." There are two classes of employees with conflicting interests – *Management* and *Operating*. Therefore, it is appropriate for these two to negotiate for their own best interest. However, "If they sink their ship, they go down together."

No corporation, association, or legal institution may engage in monopolistic or non-competitive marketplace activities, including

Certification, Accreditation, or requirements of the like. This shall not include Rating services or Apps.

The government shall facilitate independent, honest rating services and methods to aid consumers in making sound judgments about what they spend their money on. (*See 1a.*)

7a. Federal Reserve Bank

The Federal Reserve Bank (FED) shall have no responsibility for the level of economic activity. It shall regulate and provide liquidity to fractional reserve Commercial Banks. It shall maintain a Money Supply which will be expanded at five percent per annum, adjusted inversely plus or minus the rate of inflation. And it shall adjust the rate of Mandatory Savings to a greater or lesser percentage to achieve an aggregated pay-out yield on Personal Estate Fixed Investment Trust Accounts of six and one-half percent.

If, at that rate, the demand for investment capital is low and cash balances develop, the FED shall lower the Mandatory Savings Rate. If the free-market rate exceeds seven percent, the FED shall increase the Mandatory Savings Rate.

Banks shall carry all Bonds and Loans on their books at cost.

Corporate Book Value shall be the higher of Adjusted Cost or the dividend capitalized at six percent.

Until extinguished, the vast Federal debt outstanding provides a large reserve for the expansion of the money stock by monetization.

8a. Federally Chartered Mutual Savings and Loan Associations

Preamble:

The intent of this plan is to reestablish and provide equilibrium to the housing market – to match the propensity to save with the demand for housing. Forgoing immediate gratification for one's future well-being and retirement is not a natural propensity for the majority of people. Benefits from savings sufficient to induce forbearance of immediate gratification and the longstanding amortized mortgage serve individuals well and thereby society. The encouragement of ever-increasing equity in housing is a fundamental goal. Earlier experience with mutual S&Ls has taught us that requirements enumerated in this document must be immutable, for if the power to change them were to be left to legislators, those who would benefit from any change would assuredly compromise them.

The reestablished federally chartered Mutual Savings and Loan Association system, by this amendment, can eventually replace or at least diminish the role of the federal mortgage agencies: Federal Home Loan Mortgage Corporation (Freddie Mac) and Federal National Mortgage Association (Fannie Mae).

Therefore:

A new Agency will be created, which will provide initial funding, investment accounts, insurance, liquidity, and regulation exclusively to these newly federally chartered Mutual S&Ls. The new S&L system will be created by soliciting applications for charter from a specific number of contiguous zip codes in each congressional district by population. Selection of an application shall be based solely on the applicant's professional competence, experience, and community involvement. Political allegiance, race, gender, ethnicity, or age shall never be considered.

These associations must be only mutual depositories. They must never pay fixed dividends and only distribute dividends from net income.

Withdrawals must require notice and a delay of payment sufficient to differentiate S&L deposits from being considered a

"Liquid Assets" depository, consistent with available liquidity cash on a pro-rata basis. There shall be no "deposit insurance."

These associations will be prohibited from making any political donations, directly or indirectly.

These associations must never issue Certificates of Deposit.

These associations may only make new loans to new occupying owners or loans for improvement to that property, or modify loans without returning principal.

These associations may not participate in refinancing.

These S&Ls shall never make construction loans.

No Officer of the S&L may serve on the Board of Directors of any S&L.

No member of an S&L Board of Directors may serve on any other S&L Board of Directors.

S&Ls shall carry all Bonds and Loans on their books at cost.

The validity of these requirements was irrefutably established by the demise of the earlier S&L system.

9a. LAND

Preamble:

Long recognized are the three elements of production – Land, Labor, and Capital. Also, acknowledging that the genesis of nearly all land ownership is a result of conquest of some sort, hardly a concept of "good title." This solution allows the "earnings" (economic rent) of the land to flow directly to all of the consumers—an essential event if an economic system is to function close to equilibrium. This solution assures the highest use of the land over time, essential to economic growth. This will offer Native

Americans the option to participate in every aspect of the new consumer-oriented economy.

Here we are able to recognize the three elements of production, which are: the farmer's land; his and his family's labor; and the capital that includes improvements to the land—the barn, fencing, as well as livestock and equipment. With few exceptions, that which the farmer's land earns and that which the farmer's capital earns flow seamlessly to the consumer—the farmer's family. This is a significant economic relationship that this amendment codifies for all productive land.

Therefore:

All land will be designated "RESIDENTIAL" or "PRODUCTIVE."

The title to all PRODUCTIVE lands will be held in a Public Land Trust, and each adult citizen will be granted a "Life Estate" to one share held in trust. All of the lands will be subject to fifty-year land leases. The existing landowners' title will be converted to a minimum fifty-year lease and a lien equal to the depreciated value of all improvements at the end of the lease. Initially, leases will be staggered through adding two-year increments by lot. This will cause the annual auction of leases to equal two percent of the total each year in fifty years. This arrangement protects the leaseholders.

At the end of the lease, the present leaseholder will be entitled to bid an amount equal to the depreciated replacement value of the improvements. The present owner (or the lessee) holds a first lien in the amount of the depreciated replacement value of all the assets on the property. He can bid the amount of the lien or an amount to be the winning bidder, or receive a cash payment for the value he contributed to the property. Whatever amount of the auction price of the new lease that exceeds the first lien will become the Rent paid to the Land Trust for distribution to the citizenry through the CDFIT.

The Land Trust that holds title to all of the PRODUCTIVE lands shall distribute, at regular intervals, all net income to the shareholders. This will cause all of the earnings of land to flow directly to the consumers.

This will avoid the disruptive accumulations of vast windfalls resulting from land appreciation as well as mitigate the vast disparities in wealth. In this way, all the adult citizens, regardless of their wealth, will own and benefit from the ownership of all the productive land.

In the case of the farmer, this would include such things as the house, the barn, fencing, the dam of the pond, the enriched soil, and any other improvements to the raw land. After the passage of fifty years, the farm is up for public auction.

RESIDENTIAL land shall remain or become titled to the existing resident landowners.

Native Americans shall be given the option to participate fully in the benefits of the amended Constitution by subscribing to these provisions through Treaties.

FOUR PILLARS OF CIVILIZATIONAN OUTLINE FOR AN IMPROVING EVER-EVOLVING CIVILIZATION

As perceived by

The Institute of Perceptionism®

CIVILIZATION

In a world where in two lifetimes the population has exploded from one billion to eight billion, where specialization exceeds the comprehension of any single person, and human perception of this world is on a continuum where no two people view it exactly the same, in order to achieve aCivilization, a common Moral Code must exist. That which is offered here is based upon a moral code where each of the many participating individuals accepts that each individual is equally endowed with sovereignty over one's person, property, and minor children. And with very few exceptions, the mission of government is to produce, provide, protect, and preserve that sovereignty. By so doing, governmental institutions enable the sovereign citizens to interface in a free, orderly manner.

The Institute of Perceptionism® has identified four essential and interdependent pillars of civilization. These are described here and are further developed in writings at *Perceptionism.com*.

PILLAR ONE

The Nuclear Family

Wherever on Earth, the evolution of humanity over some ten thousand generations has created the Family as the foundational building block of all successful human social structures. As the human incubator, the Family has evolved to provide: the essential mothering of every human being; the psychological maturation of the well-adjusted person—a role for which the father is essential; language—the operating system for the brain; the child's socialization and civility—manners, etiquette, respect for elders, and the hierarchy; foundational education—an understanding of one's environment, economic well-being—food, clothing, shelter, healthcare, education, transportation, communication, and recreation. Also, the Nuclear Family is responsible for the basic care and compassion of its elder members. It is, inescapably, the incubator of the human being that starts before birth and lasts for some twenty to thirty years. If humanity has any value, that of birthing and nurturing a child for these earliest years is the most critically important and valuable role a woman shall ever have.

The government has not been able to successfully perform these functions. Government policies have degraded and dishonored motherhood to the destruction of the Nuclear Family. The Nuclear Family has deteriorated significantly over the last fifty years. The consequences of this have been a crumbling of the other three pillars of civilization. *PERCEPTIONISM®* offers some tools for repairing this damage. (See *Perceptionism.com.*) New institutions, such as ones that merge broken families with others in different ways, must be sought. Those who choose to raise children must be prepared emotionally, educationally, and financially to make a two-decade-long commitment to raising each child. Because of the importance of a near 24/7 commitment of the mother to the first three years of an infant's life, social and economic institutions must be styled to accommodate her productive years before and after such a commitment.

The Nuclear Families will successfully be repaired, or an ever-evolving civilization will be lost.

Suggestions for repair are disclosed below.

PILLAR TWO

A Universal Moral Code

Religious philosophy provides the human being with tools for dealing with the unknown, the death dilemma, direction, purpose, value, and a foundational moral code. Without a universally accepted and practiced moral code, there is no civilization.

Because traditional religions are taught from infancy, they become an indelible part of one's mental processing for most people—but not all. Where traditional religions have become inconsistent with human understanding of the universe for some, rejection of traditional religions has left them with a void of purpose, direction, or moral code. The default moral code becomes *Might makes right* or *the end justifies the means.*

The Institute of Perceptionism® offers a belief that "sovereignty rests with each adult human being over his/her person, property, and minor children"—not with a Governor, King, Emperor, or any government. As the sovereign people interface with one another, they must do no harm without reparation.

This tenet of self-reliance provides a simple, easily understood moral code as it requires that each sovereign individual do no harm to any other sovereign individual without appropriate reparation. This provides a moral code that is compatible with most traditional religions, as they can be practiced while honoring this tenet. This self-reliance and personal sovereignty are also the very essence of liberty and freedom.

PERCEPTIONISM®, in its entirety, not only fulfills the requirements for a universal moral code, but it also provides society with direction, purpose, and value.

(See The Human Mission)

The USA will successfully adopt a universal moral code, or it will totally lose civilization.

PILLAR THREE

Government

In a liberated world, where government is not the sovereign, sovereignty is vested in each individual and not in government; a concentration of corrosive power can be avoided. The government's primary role is to honor, protect, provide, and preserve the sovereignty of each adult human being. Sovereignty means total and absolute life-and-death authority over and responsibility for one's person, property, and minor children, with the only exceptions being those presented here.

With the implicit consent of those sovereign persons who choose to live in this civilization,

The government is granted the power and responsibility to provide and enforce the "rules of the road" and other essential rules by which sovereign individuals can peacefully, orderly, amicably, and productively interface with one another.

The government shall create institutions where sovereign individuals may have voluntary, orderly social intercourse. It shall provide courts and arbiters for the resolution of conflicts, harm, or damages that occur between and among the sovereign persons.

Persons who intentionally inflict harm or cause damage to others would be sent to behavior modification for a first or even second conviction by the judge. A third such conviction will have a mandated sentence to leave the country and never return, or be put to death. Any country that complains needs only to send a visa. A country that grants a murderer a two-million-dollar life annuity—the cost of life in prison—is too conflicted to survive.

This makes unnecessary all manner of statutes, laws, and regulations governing human conduct.

As children are commonly taught and as adults forget, "Mind your own business," not that of others, will end all manner of legislation.

Government shall enable and foster Rating Services to replace the free-market inhibiting business and professional licensing, admission-limiting associations, and accreditations. Government shall not engage in activities that can be relegated to non-government entities. For example, beyond establishing the rules of the road and the physical characteristics of the highways—with the electronic ability to toll highway use—the interstate system and its feeders can be divided into rational market segments and be placed in non-profit organizations governed by representatives of the cities, counties, and major business entities contiguous and having a local vested interest in the tolls and capacity of that segment.

This could disburse control of all the highways to over some six thousand persons employed by those representing the major stakeholders of that segment of roadway.

The administration of FAA rules can be administered by an independent entity representing and paid for by the stakeholders in air traffic control. The FDA could be replaced with a mutual insurance company where all producers and marketers of drugs underwrite all harm or damage to anyone as a result of ingesting their legally acquired drugs. This would include any side effects.

A free market for human organs would end any queuing for transplants. Such a mindset can avoid the corruption caused by concentrated power. In these ways, great power can be diffused over thousands of non-governmental persons. No government regulators may be controlled in any way by the controlee. Those who have been government employees may never be employed by controlled entities.

Government shall foster the creation of volunteer and charitable institutions to provide for matters of care and compassion.

The control of government must be limited to left-brain rational thinking. Government cannot successfully engage in right-brain matters of care and compassion as these are without limit and beyond the capacity of government and axiomatically require the involuntary participation of some to care for others—a form of slavery. Because right-brain and left-brain abilities are generally uneven among and between the genders, but equal among some, a "left-brain quotient" would be helpful in the selection of persons for government.

We have identified three evolutionary human characteristics that are incompatible with an orderly society that must be constrained by government, mitigating individual sovereignty to this extent. Also, there will be temporary encroachments as society transitions to the new, prosperous, free, and orderly society. Therefore, the sovereign persons, by remaining in this society, shall have tacitly consented to this government encroachment on their sovereignty.

First Exception

Any power over any persons or capital assets granted or obtained by an individual will influence him/her to think of how to use it to improve their personal well-being, preempting the intended purpose for one's having the power. To deal with this corrupting nature, the government shall style institutions, private or public, that govern people to function with greatly diffused and minimal power vested in any one individual by distributing it among many.

Penalties for the misuse of power must be effective.

Second Exception

Human evolutionarily designed sexual behavior is frequently in conflict with the requirements of an orderly, civilized society. Inasmuch as the Nuclear Family is paramount, irresponsible procreation shall not be permitted. In order to better assure that each human being deserves and has a sound developmental environment, the government shall issue Permits for each child to become part of a competent Nuclear Family.

Offered elsewhere in Perceptionism® writings is the following:

Irresponsible Procreation is totally incompatible with a Free-orderly society. This compromising of individual sovereignty is essential to the existence and positive evolution of humanity in a Free-orderly society. Therefore, every baby will go to its new home with two competent adults who have entered into a Family Contract to mutually and severally nurture, educate, and provide a loving, caring, developmental environment until adulthood.

Each partner of this Family Contract will have attained some minimum level of competence in order to qualify for a Permitted Child. Qualification will require certain educational achievements for the skills of a functional family, the bearing and nurturing of children, and the occupational skills necessary to support the family. At least one member shall be engaged in a productive activity that enhances the human condition, demonstrating a justification for their replacement.

Sovereign individuals entering into the Family Contract, enforced by the States, shall stipulate all of the terms and conditions of the relationship, including a procedure for dispute settlement—effectively, the Family Constitution. The Family Contract shall supersede all "marriage law."

If reverence for human life is to be preserved, conception and birth must be looked upon as life's fulfillment for the years of self-denial and forbearance necessary for the joy of achievement by competently nurturing, rearing, and educating a child for a productive role in society.

In a society where all receive sufficient income for survival, only those couples having a genetic propensity to be engaged in a productive occupation shall have the permission to procreate. This is to ensure that every newborn starts life with two competent, contractually committed parents.

To procreate without the willingness and ability to nurture and rear a child is to assume an authority without the commensurate responsibility, which is, for the newborn, the rankest form of child abuse.

If those who bear a child without a Permit or an adoptive family were to be sterilized—including the father—the pool of adopting parents would grow larger.

Inadequate parenting produces adults having psychological problems and feelings of inadequacy. The significance of the evolutionary role of each gender in the psychological development of a child will likely be discernible at a time after different family structures have been experienced in several states.

Combined with PEFITAs, Responsible Procreation will rapidly diminish social welfare programs. If Humanity has any purpose or value, Responsible Procreation is the highest achievement of most human beings.

Government shall seek ways to tame, mitigate, and civilize the human libido.

Evolution has also created a couple of human characteristics that are frequently destructive to the Nuclear Family. Promiscuity, which manifests as the pursuit of the excitement of the unconditional response, resulted from promiscuous persons procreating much more frequently than monogamous persons. Male Sexual Fixation resulted from older men who were fixated on young females procreating more frequently than those who were not. These innate characteristics cannot be denied or ignored.

Third Exception

The forbearance of current consumption and saving for future needs, while not natural to most, is essential for all to have a free, orderly, durable, ever-improving economic system. A government institution can establish a rate of return on required savings by individuals that will produce sufficient funding of the capital requirements for a rational growth rate of the economy.

The Institute of Perceptionism proposes some institutions toward achieving this end. The most significant is the Personal Estate Trust Account, which will eventually replace Social Security and the multiplicity of safety net programs, as everyone will have a sustaining passive income.

For a clearer understanding, our Proposed Amendments to the Constitution are included by reference. * The principles disclosed herein can be adopted by any government.

The USA will successfully repair its corrupt, coercive government, or it will totally lose any vestige of freedom or liberty.

PILLAR FOUR

The Economic System

The Economic System offered here is one of *PERPETUAL GROWTH.*This Economic System—*PERPETUAL GROWTH—* solves the fatal flaws of our present-day concepts.

Marxism's fatal flaw was that it required that great power be given to a few human beings who alwayINSTITUTE OF PERCEPTIONISMs became hopelessly corrupted, and the system always became dysfunctional.

Capitalism's fatal flaw was its dependence on the ever-increasing creation of fictitious assets to make up that portion of the earnings of the capital assets that did not flow back to the consumers of what the capital assets produced.

A *PERPETUAL GROWTH* economic system honors the sovereignty of each individual over one's Property. In doing so, there is no income tax. Income tax shall be replaced by excise taxes, value-added taxes, and use taxes. The role of government will shrink to a small fraction of its present-day size.

Capital is provided by personal savings invested in equity shares of the corporations, not by debt that is not secured by tangible assets. Therefore, excluding market makers, debt certification not secured by tangible assets is prohibited. Excluding debts to finance tangible improvements and new resident homeowners, it prohibits the extension of the debt burden beyond the maturity of its original indebtedness.

These measures will force a change in the ownership of capital assets from great concentrations of wealth to a much broader, inclusive ownership. Capital investment shall be supplied by forced and discretionary savings, which shall be adjusted to provide the required capital investment to maintain a durable, growing economy. This will force the sharing of the benefits of ownership.

The Institute of Perceptionism® proposes required savings instead of Social Security, placed in each individual's Personal Estate Trust Account (PEFITA). (See Perceptionism.com)

The PEFITA is a mandated savings account in which one is required to make deposits like Social Security, but it remains an earning asset that will replace Social Security and other government distributions. PEFITA must invest in a diversified equity portfolio. It can never be sold or liquidated. It can only be bequeathed to another PEFITA or gifted to a legitimate charitable non-profit organization. This assures an ever-increasing accumulation of generational wealth.

Beyond a reasonable retention of earnings for capital improvement, corporations will be required to disburse the net earnings as cash dividends. Through this system, what the capital assets earn will flow to the consumers of what the capital assets produce. There will be no buybacks, and all corporations shall have Preemption Right protection.

Institutions that enable individual ownership beyond that offer shall be encouraged. Institutions that foster personal savings, such as Credit Unions, Savings & Loans, and equity investing, are promoted.

Effective incentive plans that encourage and reward creativity, productivity, and long-term success are absolutely essential. By being rewarded with sufficient shares placed in one's PEFITA, it will be assured.

Accessible, competitive marketplaces where meticulous attention to honest, complete disclosure of risks and pricing—and matters that some buyers might find objectionable, such as goods produced with child labor—shall be assured by the government.

This means that each merchant or professional service provider shall publish and disclose the price of the service or product offered.

PERCEPTIONISM®, in its entirety, will produce an economic system which will provide humanity with a durable, constantly growing economy—avoiding booms and busts; one which distributes wealth to all of the people. Wealth is distributed consistently with nature, reflecting the wide range of human activities, human abilities, human ambition, human inclinations, and the capriciousness of good and bad fortunes; and it will fulfill the present eight human needs: food, clothing, shelter, health care, education, transportation, communication, and recreation for all.

As a disorderly society transitions to an orderly society, there will be many who need assistance with the basic needs. The government providers of assistance shall disburse only to Approved Guardians.

WARNING!

Humanity will learn of and live in a civilization of a self-reliant, left-brain-dominated population similar to that posited here, or human advancement will cease and will take on the nature of an ant hill or a bee hive. It must be recognized that, like the dog is today, each baby will be implanted with a chip that can access all of the files dedicated to that individual—unless each person is successfully sovereign over the access to one's files.

Professional Sports Teams

Professional sports teams are a product of a basic human evolutionary instinct to support tribes and tribalism. This answers the question of why thousands of people would travel miles at substantial cost to cram themselves into stadium seats to have a lesser viewing experience, endure foul weather, and be intimately exposed to crude-behaving individuals. Below the conscious level, it fulfills that need. The government allowed and enforced market impediments that enable monopoly pricing, created the immense Market Areas, inflated Market Caps, and disproportionate players' salaries. Governments created the massive stadiums. Given the understanding that "the earnings of capital assets must flow to the consumers of the assets," the ownership of sports teams appropriately should be a corporation, the shares of which can only be held by residents of the Market Area of the team, after being offered to all eligible residents through a Dutch Auction, and with a maximum shareholding of perhaps 1/10th percent of the outstanding shares. The current owners would receive the proceeds of the Dutch Auction. A continuing Bid/Ask market would be maintained by the corporation with the shareholding limit and single-share preference. Dividends would be paid only to persons having addresses in the prescribed Market Area. This would complement the public ownership of the stadia

Made in the USA
Monee, IL
04 November 2026

67386632R00138